Engineering, Your Career

Engineering, Your Career

By Howard P. Rosenof

ISBN 979-8-9851706-0-3

Library of Congress Control Number: 2022908111

This publication contains the opinions and ideas of its author. It is sold and otherwise distributed with the understanding that the author is not rendering legal, tax, investment, financial, educational, or other professional advice or services. If the reader requires such advice or services, a competent professional should be consulted. The strategies outlined in this book may not be suitable for every individual, and are not guaranteed or warranted to produce any particular results.

Mention of specific websites or authorities in this book does not imply endorsement by the author, nor does it imply that such websites or authorities have endorsed the author. Internet addresses given in the book were accurate before the time it went to press. The author is not responsible for changes to third party websites.

About the Author

Howard Rosenof specialized in control systems for power generation, energy management and manufacturing. He is recognized internationally for his contributions to the control of batch chemical manufacturing processes, as the first American to receive the Dr. Guido Carlo-Stella Award from the World Batch Forum. He is also known for his work in applying artificial intelligence to problems in industrial automation. He holds a Bachelor of Science degree from the College of Engineering of Cornell University, and an MSEE from Northeastern University. He is a licensed Professional Engineer in Massachusetts.

Also by Howard Rosenof:

Batch Process Automation, Theory and Practice
1987, co-author with Asish Ghosh

Utility and Independent Power: Concept for a New Millennium
(F. William Payne, Ed.)
1997, chapter author

Table of Contents

Forward

Every child looks for role models.

For me, I had two very strong but opposite ones. My family spanned the classic "left brain – right brain" spectrum. My sister Susan, a talented artist, represented the right brain. My brother Howard, the author of this book, exemplified the left brain. If there is any doubt, refer to his explanation of pronouns in the introduction to this book. Of course, there is always some overlap, but in general the distinction is real.

I always felt I had to pick one side. I was five years old when my brother left to pursue a degree in Engineering. Until then, my only reference point for "engineers" had to do with trains. So, I was enamored with the idea that you could go to school to learn how to drive trains. What five-year-old boy could resist?

As I grew up, I began to think more like an engineer than like my creative sister. Despite being a musician, as well as an avid fan of the arts and architecture, I was drawn to the problem-solving prowess of engineers. While my brother may have a much more eloquent way to express it throughout this book, an engineer's purpose is to solve for "x". And what can be a greater gift to society than problem solving? Doctors, lawyers, teachers, business people, construction workers, politicians –even children on playgrounds problem solve. After all, what do doctors do when faced with sick patients? They find the variables, apply the equation and formula, and solve for "x". See Mom, I could have been a doctor!

Throughout the book my brother explores the topic of the outward-facing perception of engineers—how they are perceived by others. As Howard states, the term "engineer" is very loosely defined. To be a physician, one must earn a well-defined degree accredited by one of a very few specific organizations, then complete additional practical training under close supervision. Yet the sound man for Led Zeppelin is considered an engineer in the same sense as WSP, the structural designer for the Freedom Tower in New York city, is an engineer. Could this be the reason an engineer is not thought of in the same way as a doctor or lawyer? I will let my brother further explore that topic.

So why do engineers sometimes lose their societal rewards either in status or money? Seemingly, the smartest folks in the room are often relegated to their cubicles, slaving over spreadsheets and algorithms for their entire careers.

My personal theory presumes that the ability to communicate effectively is the key differentiator. I believe one starts their career getting paid for what they do, then paid for what they know, then paid for whom they know, and then finally for the ability to communicate a strategic vision that propels a team or entire organization to a high level of success. Some of the smartest people I know – many of whom are engineers – never get past the first level. While Howard will explore parts of this continuum throughout the book, I ask that you focus on the latter. As an example, I am a General Contractor by trade. So, while interacting with engineers is a large part of my job, easy, concise and communications with all the stakeholders on a project is my key to success. For a time, I was working for an international engineering firm as a project manager. One of my reports was a graduate of a very prestigious Massachusetts technical institute (you can guess). I was trying to understand why two pieces of equipment would not interface, I was lucky enough to receive a 10-minute lecture about Bernoulli's Equation and how I just could not be expected to get it. I replied, "does the blue box talk to the red box"? Dejected, my engineer report looked at the floor and answered, "Well yes – I guess you can look at it that way".

Albert Einstein once said, "You do not really understand something unless you can explain it to your grandmother." No offense to anyone's grandmother.

My advice to everyone reading this book is to be a little bit more like Albert; call your grandmother once in a while, and enjoy the book.

David Rosenof, State of Florida Certified General Contractor (CGC), Project Management Professional (PMP), Leadership in Energy and Environmental Design Accredited Professional (LEED-AP), MBA and Howard's baby brother.

Introduction

This book is many years in the making. As an engineering student, I had roommates and friends in other majors who went on to study law and medicine. As young adults, our situations were not identical, but not radically different either. As time went on, our trajectories diverged – a law student became a partner, a medical student became a prominent physician, and the engineers I knew as students either left the field, moved to general management, or rather early in their careers experienced stagnation. Without any exception that I can recall, among my contemporaries and near-contemporaries who stayed in engineering, every one, myself included, has experienced intervals of involuntary unemployment, sometimes exacerbated by non-compete "agreements" that outlived their jobs. This led to a decades-long interest in understanding what was different about us, America's engineers. I looked into the way we're educated, our place in society, and the issues that many of us experience in our careers. Given our nation's emphasis on education in science, technology and mathematics (STEM) and the importance of engineering generally, I hope that my attempts to find answers to these questions will be of value to American engineers and students considering engineering careers, as well as others who may be interested.

In this book I ask "who are the engineers?" The answer is not as obvious as one might think, or as it would be for physicians and lawyers, among others, and the answer changes over time. I got my first degree in engineering without ever being told that there was such a thing as an engineering license. Are you an engineer if you're not licensed as an engineer? Are you an engineer if you have a job title that says "engineer" but don't have a four-year degree in engineering? Are you an engineer (a "software engineer") if you've studied computer science but never had to take a course in thermodynamics or strength of materials? I propose answers to such questions.

I consider the role that we engineers play in society. While some hold licenses as Professional Engineers, most people engaged in the practice of engineering do not. I ask whether engineering is a profession or something else — perhaps a trade or some other form of occupation. Are engineers, with or without licenses, considered professionals? I look into claims, typically by employers and government officials, that there is a persistent shortage of engineers, and if there is one, whether it arose in some organic fashion or can be traced to educational policies or the way engineers are treated by employers.

I then discuss the process of becoming an engineer. Today's default path is through four years of post-secondary education, but this hasn't always been the case; engineers used to learn by apprenticeship, and there were engineers long before there were engineering schools. I give you some hints that may help those considering becoming an engineer find the best path.

Finally, I consider issues that many engineers encounter as they move through their careers, such as salary compression and   age discrimination.

This book's intended audience is people much like myself – "working engineers" as I was for most of my career, as well as people aspiring to become such. "Working engineers" are engaged primarily in the practice of engineering, rather than teaching or management, particularly the management of non- engineering staff. In essence, I've endeavored to *write* the book that I wish I had *read* a half-century ago.

You will read some comments about the engineering profession which you are likely to perceive as negative. Does this imply that my wish is to discourage young people from entering the profession, and older people from remaining? No; it's the opposite. Our American society needs a strong engineering profession to help us maintain our first-world life-styles in a sustainable way, allow others to attain similar lifestyles, and to keep our adversaries at bay. Even those who may find this language "corny" are doubtless at this moment enjoying the freedom and prosperity that our system enables. Engineering has helped enable upward mobility, in that an individual can qualify for a career after only four years of post-secondary education without the direct expense and opportunity cost required to go on to a graduate "professional school".

The world population is now about 7.8 billion and growing by tens of millions per year. It's engineers, in conjunction with scientists, investors and others, whose services will be necessary to allow our society to continue to enable the lifestyle we enjoy, and permit others to achieve it, by making the most efficient and benign use of available resources. We engineers have a lot to do, which is all the more reason for society as a whole to allow the profession to become more attractive, not only for those who want to have their "tickets punched" on their way to senior management but those considering making engineering their real career. To make that happen, though, engineering has to become more competitive with some of the other options and opportunities that are available to smart young people. You'll learn what I mean by reading this book.

I am also not suggesting that our mostly-capitalist economic system should be replaced by anything else; replacements have long been proposed but never, in my opinion, proven. Nevertheless our system works best when opposing interests meet in the marketplace and converge on policies and solutions that all can live with. When one buys or sells a house, prices reflect the momentary balance of supply and demand but for the most part there is an established process that is generally fair to both parties. In workplaces covered by collective bargaining agreements, labor and management negotiate working conditions including compensation. But I will argue that the circumstances under which most engineers work in the United States reflect an overwhelming imbalance of power, including influence over government, enjoyed by employers.

I don't say much about questions that are common to employees or students generally – for example, how to pay for college. There is no shortage of resources available to help answer these concerns. I also don't discuss issues that may apply primarily to one or another subgroup of current or prospective engineers. However one or more of these resources may apply to you: The National Academy of Engineering (www.nae.edu) provides an "engineer girl" website. The National Academies Press (www.nap.edu) offers such titles as "Seeking Solutions: Maximizing American Talent by Advancing Women of Color in Academia" and "Promising Practices for Advancing the Underrepresentation of Women in Science, Engineering and Medicine". A variety of identity-based organizations are available, including among others the National Society of Black Engineers (nsbe.org), Society of Hispanic Professional Engineers (shpe.org), the American Indian Science and Engineering Society (aises.org), the Society of Women Engineers (swe.org) and the National Organization of Gay and Lesbian Scientists and Technical Professionals (noglstp.org).

A few words about words: In the old days, the English language gave us three nominative-case pronouns we could use for people – he, she and they. To refer to a person who might be a male or female by his or her role, e.g. a police office, teacher or engineer, one would use "he or she", the more common version, or "she or he". We simply don't have one pronoun we can use to equally denote both possibilities; "they" is traditionally plural so is used to refer to two or more individuals. The "singular they" is gaining popularity, although it can create ambiguity. Some organizations are publishing pronoun dictionaries to account for a larger variety of contemporary possibilities, and some states are allowing at least one gender other than male or female to be claimed in official documents such as driver's licenses. Any author who wishes to refer to an individual by role now has a stylistic problem; even using "he or she" or the reverse, which doesn't make for the easiest reading experience when used too frequently, could be seen to unnecessarily exclude some. In the interest of writing something comfortably readable, without intent to exclude or offend, and making reference to the Dictionary Act of 1871[*], I will use the pronoun "he" for engineers and "she" for lawyers and others. I do this only to write readable text, with full recognition that there are female engineers and male lawyers, and understanding that at any given time there are likely people in all professions who place themselves into neither gender or who who are in the process of transition from one to another.

In our modern world, a few disclaimers will be in order:

This book does not contain legal advice. Legal advice is given by properly-licensed attorneys. I am not an attorney. If you have a legal question or issue, contact a properly-licensed attorney. When I write about the law and legal issues, I do so only as an interested layman.

Comments on investments, financial strategy and taxation represent only my beliefs and opinions. With respect to such matters, I write only as a layman. Readers must conduct their own due diligence and obtain professional advice in whatever combination they deem appropriate.

Advice contained herein is based on my own experiences. Every engineer, employer and situation is different.

Settled disputes rarely if ever include an admission of guilt or other wrongdoing. No such admission should be inferred from any reference to a dispute having been settled.

Assertions of fact are based on source material that I believe to be accurate, but for which no guarantee is made.

The author accepts no liability for any losses or damages (whether direct, indirect, special, consequential or otherwise) arising out of errors or omissions contained herein.

I am grateful to those who offered comments on early drafts. These include Sue Rosenof, who also designed the cover, Zachary Rosenof, Nathan Aronow, Peter Thelin, Carl Mikkelsen, Rachel Gordon, and Judi Roth. I particularly wish to thank my wife Jane both for her support through the years it took me to get this project done, and for her detailed review of the manuscript.

Attorney Carol Schurr Levin assisted me with legal questions pertaining to this project.

---

[*] "...words importing the masculine gender include the feminine as well"

Chapter 1: The Engineer

What do engineers do?

The practice of engineering incorporates several components, which are present in individual jobs to varying degrees.

Design: This is the core of engineering for many of us, and is probably the first thing that non-engineers associate with the field. Individual components of a system must often first be designed, and then the components assembled into the system. The process may comprise many levels of integration. Take the design of an automobile as a simplified example. Engineers collaborate with product designers (a slightly different use of the word "design") to define a product which is aesthetically pleasing while meeting certain specifications—some regulatory, like crashworthiness, and some commercial, like load-carrying capacity. Then engineers have to specify the car's components which, if not already available, engineers must design. So let's say the car is to be an all-electric, making battery capacity critical. The required capacity will depend on the car's weight, its desired range, and its targeted cost. If the right battery isn't already available, the car manufacturer will have to collaborate with a battery manufacturer. The battery manufacturer in turn will have to design a new battery. And so on. Also, in addition to designing "things" and "stuff," engineers design processes, e.g. for manufacturing.

In this example, it is not within the engineer's scope of responsibility to determine, for example, how much trunk space the car should offer. This is a decision made by marketing or some other business function, based on current and predicted demand, the market price that is expected to optimize profit, and so on. The Fukushima nuclear disaster in Japan in March of 2011 gives us a real-world example – the specification for the site called for it to withstand a tsunami 5.7 meters high (above sea level) but the actual tsunami height that day was about 15 meters.[1] Even the best engineering can prove to be inadequate when it is performed to comply with an inadequate specification.

Of course, there's no reason why an engineer can't himself also define a product or process to be designed. These engineers then often become entrepreneurs, first working individually or in a small group, then attracting investments to further develop and commercialize a product.

Research: Engineers in academia or research institutions often work alongside scientists to advance scientific knowledge. Practicing engineers are also engaged in research, more often to adapt scientific findings or available technologies to new applications. This can be termed industrial research, especially when engaged in by large teams. "Research" may also mean poring through manufacturers' websites to find a product that it likely to meet the engineer's requirements.

Thomas Edison is credited with creating industrial research and development, after having been an inventor first working alone and then with a small staff. His laboratory in Menlo Park, New Jersey tested thousands of prospective filament materials until he found that carbonized cotton sewing thread could serve as a filament for a practical electric light bulb. As is not uncommon, Edison's research had to go beyond existing scientific knowledge in that he wasn't trying to generate theories to explain observed phenomena, but simply looking for things that would work, no matter how.

Development: Development is closely related to design but carries a sense of taking information obtained through research and commercializing it.

Once Edison had a phenomenon (incandescence) and a material (cotton thread) that he could exploit, his bulb needed further development to be commercially successful and then to try to keep ahead of competitors. Filaments evolved in several steps from the original cotton thread to drawn tungsten, through innovation both in Edison's labs and elsewhere. Today, incandescent lighting for general use has been all but replaced by light-emitting diodes – LED's – in an example of disruptive technology.

Training, Education and Communication: Engineers as faculty help train new generations of engineers. Engineering consultants and engineers in industry often also impart information to others – their fellow engineers, other personnel such as assemblers and technicians, customers, managers, and regulators. Information is conveyed in many ways, ranging from simple reports to training courses and books. Some engineers focus exclusively on helping customers select and then use their employers' products.

Manufacturing, Operating and Construction Support: Practices vary among industries, but engineers' responsibilities don't end when a design is released for construction or a product goes into production. Engineers help solve problems that come up once their designs are implemented, and some problems don't emerge until the design's results have been put into regular use. Engineers may be called on to assist with testing and other tasks ordinarily performed by other personnel, and help direct construction activities.

As but one example, when the yield of a manufacturing process begins to drop (that is, the process produces more waste material or a higher reject rate than it should), if production personnel can't resolve the problem it is likely that an engineer will be called in.

Sales: Engineers support their employers' sales activities, particularly when technical products are sold to other businesses or governments rather than to consumers. When prospective customers issue requests for proposals, often for work that requires customization, engineers often help prepare the proposals. Many engineers move into full-time sales roles as their careers progress.

Leadership and Management: Non-management engineers often serve in lead roles, where they do not have supervisory responsibilities such as for salary management, but do coordinate their teams and take responsibility for work quality, schedule compliance and budgets. Engineers are often promoted into management, where especially in supervisory-level roles they continue to rely on their technical knowledge to help guide the people who report to them.

Who is an engineer (and who isn't)?

For the purpose of this book, an engineer is an individual who applies mathematics and scientific knowledge (physics, chemistry, properties of materials and increasingly biology) to design or improve structures, public works, products and/or processes. In the words of the US Bureau of Labor Statistics (BLS),[2] an engineer:

> Performs professional work in research, development, design, testing, analysis, production, construction, maintenance, operation, planning, survey, estimating, application, or standardization of engineering facilities, systems, structures, processes, equipment, devices, or materials, requiring knowledge of the science and art by which materials, natural resources, and power are made useful. Work typically requires a B.S. degree in engineering or, in rare instances, equivalent education

and experience combined.

Excluded are:

a. safety engineers;[*]

b. sales engineers;

c. engineers whose primary responsibility is to be in charge of nonprofessional maintenance work;

d. engineers in charge of programs so extensive and complex (for example, consisting of research and development on a variety of complex products or systems with numerous novel components) that one or more subordinate supervisory engineers are performing at level 8;

e. individuals whose decisions have direct and substantial effect on setting policy for the organization (included, however, are supervisors deciding the "kind and extent of engineering and related programs" within broad guidelines set at higher levels); and

f. individual researchers and consultants who are recognized as national and/or international authorities and scientific leaders in very broad areas of scientific interest and investigation.[†]

Ordinarily an engineer will have completed at least an accredited, four-year post-secondary program in a field of engineering. It is said

---

[*]   At least one state, Massachusetts, licenses individuals as safety engineers.
[†]   Even the official BLS definition admits there are individuals given the title "engineer" who should not be recognized as such.

that the French were the first to require rigorous formal training (i.e., rather than apprenticeship) for their engineers, and that their word "ingenieur" is based on the Latin "*ingenium*", which is also the basis for the English word "ingenuity". The word "engine" is also related, its early use implying that the engine was the result of human ingenuity. Then at some point the word "engineer" in English became associated with operating, not just creating, ingenious things. Because of this linguistic turn of events and because the title Professional Engineer is legally regulated by the states, through powers that the federal BLS definition does not recognize, there is no clear definition in the American English language of who an engineer is. In my opinion, this ambiguity is at the heart of the problem faced by the profession and its members. Consider also that physicians, for example, no matter the balance in their careers between clinical practice, research and administration, don't stop considering themselves physicians; they advance within, rather than out of, their professions.

This book's definition of "engineer" generally follows the BLS definition. For the purposes of this book, "engineer" includes practitioners in the major areas of civil, mechanical, electrical, structural, chemical and industrial engineering, as well as their sub-specialties like electronics engineering; newer branches like biomedical engineering, and the hybrids like manufacturing engineering and environmental engineering. It excludes people who only have such credentials as Microsoft Certified Systems Engineer*, who need not be formally trained in science and mathematics. Same for Certified Quality Engineer, Reliability Engineer, and Software Quality Engineer credentials as granted by the American Society for Quality.[3] A cost engineer might be the holder of a four-year degree in engineering, but the function doesn't necessarily require it and it's not a recognized branch of engineering.

A stationary engineer is a member of a trade with responsibility for operating (not designing) boilers and other major items of equipment. ~~This traderequires~~ licensure in many jurisdictions, and holders of engineering degrees who wish to obtain such a license may in some cases be

*Renamed Microsoft Certified Solutions Expert in 2012

given credit toward its required minimum years of experience. Related titles are operating engineer and power engineer, with operating engineers being primarily employed in construction as operators of complex equipment. A chief engineer may be a senior power engineer with management responsibility. Power engineer is in some cases synonymous with operating engineer, but electrical and mechanical engineers working in electric power generation and other heavy industry may also refer to themselves as power engineers. Locomotive engineers have a role similar to that of stationary engineers but, well, they're not stationary. While there's nothing to prevent an engineer as recognized herein from also joining a trade, none of these trades is considered an engineering occupation.

A hotel may send up an "engineer" to unclog a toilet. A major financial services firm advertised at one time for a lead operations engineer position that required a high school diploma and an electrician's license. In the United Kingdom an "engineer" may be assigned to fix a home heating system.

A financial engineer is often a person who trained as an engineer and then moved into financial management on the basis of an MBA or an advanced degree in economics. This individual is not working as an engineer.

Many job titles include the word engineer but do not plainly demarcate engineering work as the term is applied here. Our definition of "engineer" does not automatically include people with titles like field engineer, commissioning engineer and support engineer, who in many cases are technicians and not required to have four-year engineering degrees. Employers are of course free to require four-year degrees as qualification for these titles if they wish. "Engineer" as used here excludes those in various industries who have titles like engineering technician and technical engineer, as for these also a requirement for a four-year engineering degree is at the discretion of the employer. However, an engineering technologist is often a graduate of a four-year program in engineering technology, and a geotechnical engineer (or geosystems engineer) is indeed an engineer.

An aerospace engineer is an engineer, but in the absence of other credentials a flight engineer is not. A military combat engineer is not necessarily an engineer, although graduate engineers may well be leaders and members of combat engineering units. An audio engineer who designs audio equipment is generally an engineer, with a degree, but to be a sound engineer who manages the sound system at a public performance, or a studio engineer who operates a recording, television or similar studio, an engineering degree is not required. A marine engineer, particularly a marine engineering officer, usually has the equivalent of a four-year university education in engineering although the training also includes technical skills like welding. An architectural engineer is usually an engineer (except for those who hold only certificates and not four-year degrees) as is a structural engineer. A building engineer probably maintains a building's HVAC (heating, ventilation and air conditioning) equipment, but a Green Building Engineer holds a certification granted by the Association of Energy Engineers only to licensed Professional Engineers who meet additional requirements. Ceramic(s), metallurgical, materials and mining engineers are engineers, even though some states may not recognize these as professional engineering disciplines.

Once on a local news broadcast I saw an interview with a trash collector who was identified, in a crawler below the image, as a sanitation engineer. Trash collection is an honorable occupation and makes a greater contribution to public health than many realize (imagine an urban area without it) but it is not a branch of engineering. Sanitary engineering, however, is a well-established specialty within civil and environmental engineering. It is concerned primarily with wastewater facilities and also makes an immense contribution to public health. Civil and environmental engineers may also, of course, work in solid waste management.

Agricultural engineering is an established engineering discipline although it may be taught in an agriculture school rather than in an engineering school.

Plumbing engineering is not a recognized engineering discipline, although mechanical engineers may design plumbing systems and may then be reasonably referred to as plumbing engineers.

Titles like project engineer and test engineer refer more to an individual's function within an organization than to the level of training and professional credentialing required for the job, just as a project manager doesn't necessarily have a degree in management. It would be clearer if these prefixed titles were always restricted to degree-holding engineers, but they are not, so the titles don't tell us whether their holders must be engineers. The title application (or applications) engineer may be used in a similar fashion to test engineer, but the term applied engineering has also been allowed for four-year programs in what is more commonly called engineering technology.

Associate engineer may be a corporate title but is more commonly a reference to the holder of a two-year degree with a name like "Associate in Engineering". These degrees are typically given by community colleges via programs that are generally the equivalent of the first two years of a four-year program.

A Professional Engineer holds one or more state licenses to practice engineering. Without the capitalization, the title "professional engineer" is ambiguous and should not be used. In particular, those without professional licenses should not claim titles that could leave themselves open to allegations of misrepresentation. The term "engineering professional" is similarly imprecise. (Note that "medical professional" usually includes people who are not physicians.) Professional Engineers can use their proper titles and others should choose clearer terms. Professional Engineers may identify themselves with the initials PE (or P.E.) after their names. Professional Engineers are often advised to specify on their physical and virtual business cards the state(s) in which they are licensed, lest they be accused of offering engineering services in jurisdictions which have not licensed them to do so.

The term Professional Engineer is granted in the United States by state governments (and their equivalents in such jurisdictions as Puerto Rico), but the registered trademark "Professional Traffic Operations Engineer" (PTOE) is a certification given by a private organization, the Transportation Professional Certification Board, Inc.[4] The Certification Board indicates that Professional Engineer licensure is a requirement for the PTOE certification in the US, Canada or any other jurisdiction in which the engineering profession is regulated.

As another example of a non-governmental organization granting certifications that include the word Engineer, the International Association for Radio, Telecommunications and Electromagnetics (iNARTE) conducts a range of activities in areas around telecommunications, among other things administering U.S. commercial radio operator licensing tests on behalf of the Federal Communications Commission. Such licenses are required, for example, to operate broadcast television stations. iNARTE offers an EMC [Electromagnetic Compatibility] Design Engineer Certificate[5] as well as other credentials that include the word "engineer". There are requirements for education and experience that generally parallel those for Professional Engineer licensure, but in contrast to the policies applied for PTOE certification, licensure is not required. Therefore an individual so certified is not necessarily a state-licensed Professional Engineer.

The ambiguity in titles extends to company names. "Smith Engineering" is probably a consulting engineering company once or presently headed by a Professional Engineer named Smith, but "Auto Engineering" is usually a car dealership and "Jones Heating and Cooling Engineering" is probably, but not necessarily, a contractor and not an engineering firm. (To be clear, these names are all contrived for use as examples. They do not refer to any actual organizations or individuals.) In some jurisdictions engineering companies must be licensed in addition to employing licensed personnel. To quickly get a sense of how aggressively "engineer" titles are enforced where you live, look for mechanical contractors in your area and see how many of their names include the term "engineering".

What about people who received engineering degrees but immediately went off to business school, law school or the like? On the basis that the usual path to a professional engineering license requires some years of experience working as a degree-holding engineer, I do not consider these business managers, lawyers, etc. to be engineers, but this one is arguable. Once someone has earned an accredited degree in engineering and has worked as an engineer for a while, I have no objection to his describing himself as an engineer but not a Professional Engineer unless currently licensed. Interestingly though, in my experience, once they've changed fields, such non-practicing engineers are often loath to describe themselves as such unless to support a point they want to make, e.g., "I'm an engineer so I know how to interpret the data".

One friend, who has many years of success in engineering but wouldn't quite qualify according to the definitions above, noted that "I may suffer the consequences of being an engineer without actually being one". These definitions are meant more to define the term as used in the rest of the book, than to arbitrarily exclude anyone from the field.

Can a chemist or a physicist be an engineer?

Engineering practice involves a balance between the white-collar world of mathematics and scientific inquiry on the one hand, and the blue-collar world of making things actually work on the other hand. In that world schedule delays cost money, batteries run out of power at inconvenient times and corroded pipes have to be dug up. It's generally accepted that practical skills can be more easily learned on the job than can math and science. Consequently a four-year education in physics, for example, is considered reasonable preparation for many engineering jobs. State laws may vary in their acceptance of related but non-engineering degrees as qualifications for professional engineering licenses. Texas, for example, considers "related science degrees" as equivalent to non-accredited engineering degrees, accepting these degrees but requiring additional experience for licensure.[6]

The field of engineering physics combines science and its practical uses. States generally don't recognize engineering physics as a branch of engineering, but some license nuclear engineers, in which case they may recognize these degrees without requiring additional experience to make up for a perceived educational deficiency.

Are Engineering Technology graduates engineers?

Starting around 1967, the traditional distinction between an engineer with at minimum a four-year post-secondary education, and a technician with typically two years of such training, was complicated by the advent of engineering technology programs at the baccalaureate level. Today it's even possible to get a masters degree in engineering technology.

Engineering technology programs de-emphasize but certainly don't eliminate the engineering program's math and science courses in favor of more lab work. States vary in their acceptance of engineering technology degrees as prerequisites for professional engineering licensure; check your state's licensing board if this question applies to you. Commonly, states do accept these degrees but require additional years of experience.

Walter W. Buchanan of Texas A&M University explains that the four-year engineering technology degree came as a result of the "Sputnik Scare" some years earlier in 1957, which prompted traditional engineering programs to reduce their emphasis on lab work in favor of a greater emphasis on theory.[7] By this view, engineering technology programs in the United States are closer to pre-Sputnik engineering curricula than are contemporary engineering programs. In fact, the usual definitions of engineering as the "application of scientific and mathematical principles..." or some such, are only partly true. Early engineering was empirical, and in many cases came well ahead of scientific explanations of the physical phenomena they employed. Ancient civilizations built roads and canals by trial-and-error, not theory. Hero of Alexandria built his steam engine about 1700 years before the laws of thermodynamics were finalized. There is still an empirical element to engineering today – we still test our decisions, in a variety of ways according to context. So engineering technology can be seen as a movement to reclaim the early history of the field. Engineering technology programs are offered mostly at public institutions, and typically not on their flagship campuses.

Engineering work requires a range of skills, with different positions emphasizing different skills. Most engineers don't primarily do research, just as most physicians don't. Look at a typical description for an engineering job and you'll see tasks with elements of research, testing, design, programming, documentation, teaching, troubleshooting, and so on. Engineering technology graduates can often perform well in jobs more commonly filled by traditionally-trained engineering graduates.

In light of the above, my definition of engineering does in fact include graduates of engineering technology programs. We can view any undergraduate engineering education program as lying along an axis marked as "theoretical versus practical". Science programs – physics, chemistry and biology – dominate at one extreme with an emphasis on knowledge and research. Engineering technology programs are at the other end focusing on practical uses of this knowledge, and traditional engineering programs in the middle. There is, of course, overlap. These three approaches are all important, in different ways, and all should be considered legitimate paths to reaching recognition as an engineer. With respect to licensure, however, the profession does involve a balance between theory and practice. It's reasonable, therefore, for graduates of programs at the extremes of this axis to be required to document more years of professional experience than graduates of traditional programs.

Do non-accredited degree programs produce engineers?

Not all engineering programs are created equal. The most rigorous are accredited by the Accreditation Board for Engineering and Technology (www.abet.org). ABET also accredits engineering technology and other programs, but the standards are different and the accreditations are not equivalent. It can't be assumed a "degree in engineering", even from a prestigious university, is accredited. For example Yale University offers no fewer than three flavors of engineering degrees – the Bachelor of Arts in Engineering Sciences, the Bachelor of Science in Engineering Sciences, and the ABET-accredited Bachelor of Science. Only the third, as the name implies, is fully accredited. While accredited programs are required to meet the standards that apply to them, educational administrators have a great deal of flexibility in designing non-accredited programs.

Be aware also that there are two types of accreditation. Regional accreditation agencies, like the North Central Association of Colleges and Schools, accredit institutions as a whole, not specialized programs such as engineering. ABET is the organization that accredits engineering and engineering technology programs across the United States.

It is my opinion that only ABET-accredited engineering and engineering technology degrees are appropriate for those who wish to practice in the traditional fields of engineering after only a four-year college program. Non-accredited programs may be fine for students planning to move on to other professions. However, those with non-accredited degrees may find themselves at a disadvantage when seeking employment or graduate education in engineering, or Professional Engineer licensure. For the purposes of this book, references to US engineers can generally be presumed to apply to individuals holding at least one ABET-accredited engineering degree, an accredited engineering technology degree or a recognized foreign equivalent, or Professional Engineer licensure achieved through some other path.

ABET will not accredit a program that doesn't have graduates, so it can take years for a new program to become accredited. Accreditation may be made retroactive, however.

Is Software Engineering a branch of engineering?

"Software Engineer" is tricky. Many individuals with traditional engineering qualifications including professional licenses have moved to careers in software application development and other specialties within information technology, and they continue to view themselves as engineers. But it is a presumption of engineering that engineers deal with the physical world; are you still an engineer if you spend all of your working time writing code, say, to manage websites? Are you any kind of engineer if you haven't had to study years of college-level math to get your degree? Many would say that the answer is no.

There are a number of universities offering degrees like Master of Software Engineering[8] and Master of Science in Software Engineering.[9] For the most part completion of a traditional engineering education at the baccalaureate level is not required for admission into these programs; only a degree in an "appropriate discipline" is needed. Since one can earn even an advanced degree in software engineering without ever having studied the fundamentals of engineering, this author is not prepared to give automatic recognition to graduates of these programs as "engineers", any more than an individual with a doctorate in pharmacology should be licensed to practice as a physician. Graduates of software engineering programs that *do* cover the same principles of science and mathematics that more traditional engineers have to learn are, in my view, entitled to an exception to the exception, and to recognition as true engineers.

With respect to licensure, the National Council of Examiners for Engineering and Surveying (NCEES) offers an electrical engineering examination specialized in electrical and computer engineering, but this is only indirectly related to software engineering.[10] Texas declared its intention to provide professional licensure for software engineers and in 1998 granted the first PE license in software engineering.[11] In 2013 the NCEES introduced a test in the area of software engineering but then announced that it was discontinuing the test after April 2019 due to low interest among candidates (19 registrants for the 2018 exam).[12] Individual states will still be able to design their own tests but it appears unlikely that many will do so, given the low level of interest among prospective registrants.

There may be be some unintended consequences to the recognition of software engineering as a branch of engineering. Many software engineers are freelancers, operating as contractors or consultants. To the extent that software engineering becomes formally recognized as an engineering specialty, practitioners in the field could find that they have to be registered as Professional Engineers in order to offer their services to the public, if "the public" is defined to include some businesses. To avoid such a situation people in this category may do better to refer to themselves as "application developers" or with some other title that does not include "engineering".

Notwithstanding the question of whether software engineering is a branch of engineering, software engineers bring to their vocation a level of preparation generally similar to that of traditional engineers. They also face many of the same career-management issues, considered later in this book, that are faced by traditional engineers.

Conclusion

I don't claim that my definition of engineering and my rules for who should and should not be referred to as an engineer are the only valid ones possible. The federal government disagrees with state governments, and the states with one another, about who deserves the title.

Individual educational institutions, employers, labor unions, journalists, software developers and engineers collectively and individually may all have their own reasonable but different answers to the questions of who is an engineer, what kind of work is engineering work, and what fields constitute the branches of engineering.

A lot depends on what, adjective, if any, is placed in front of "engineer". Lawyers don't have this problem; real-estate lawyers and bankruptcy lawyers are lawyers, and we all understand that a "jailhouse lawyer" isn't really. Physicians don't have this problem either – medical doctors in the United States have long since agreed to include osteopathic physicians as members of the same profession. There is also rarely confusion caused by the presence of individuals with doctorates in other fields; e.g. physical therapists, clinical psychologists and holders of doctoral degrees unrelated to medicine are not routinely confused with medical doctors.

There is simply not an unambiguous boundary that defines who is a member of our profession and who is not. We as engineers have never been able to establish such a boundary, and it's understandable that society as a whole can't either. This situation does not benefit America's engineers, who have never been able to develop the high levels of cohesion that allow other professions to more effectively pursue their collective interests.

Where do we come from?

The U.S. government, specifically the National Science Foundation, keeps careful track of the demographics of engineering students and of practicing engineers.[13] Their reports can tell you, for example, how many people working as engineers and scientists in the United States have no degrees in engineering or science (about a million). For the most part I will leave the review of these statistics as an exercise for the reader, and move to a related set of questions, which I find more interesting: Who becomes an engineer? Why do they become engineers? Who stays in engineering and who leaves the field? And finally, can we discern any differences in interests and motivation between the engineers who grew up here, and those from overseas? To answer such questions, there are only a few studies and minimal statistics, so except as noted I rely upon my 45-plus years of participating in and observing the field.

Engineering doesn't provide the prestige and long-term income of some other fields like law and medicine. Unlike those, however, entry into the engineering field doesn't require years of post-graduate education (or for some, apparently, any engineering education at all). And starting salaries tend to be high compared to those of other fields that require only a four-year degree of most entrants; all of the top 10 majors with the highest starting salaries are engineering fields.[14] This can be an attractive combination for a graduating high-school student of modest means. Furthermore, students from middle-class, blue-collar families are likely to have had some exposure to technology through their parents, who might be farmers (farms were the source of many of America's early technology innovators), auto mechanics, heating technicians, and so on. Such young people are likely to see engineering as a natural and affordable next step up the educational and social ladder.

The children of engineers constitute another group of young people who seem to be disproportionately represented among people who become engineers. Both groups appear to gravitate toward engineering for its own sake; that is, they study engineering to become engineers.

There's another group, though, that looks at engineering school as a step on the way to a different goal, such as entrepreneurship or corporate management. These in my experience tend to be the children of corporate executives, successful entrepreneurs, lawyers and so on.

Some actually work as engineers after getting their first degrees, and others go directly to business school or other postgraduate training outside of engineering. An undergraduate education in engineering gives them analytical skills, the credibility of having completed a difficult course of study, and a knowledge of some of the vocabulary and technologies used in the economy. Commonly they don't intend to make a career in traditional engineering, reaching that decision either before they graduate from college or after a short time working.

Engineers from overseas, in my observation, are often in the second category, not planning to spend too much time doing hands-on engineering. In the context of their own societies they are often from families that are better off than those that American-born engineers grow up in. I believe that this is because developing economies are just that – developing. "Developing" means that their countries are creating industries and building infrastructure, whereas in the developed world, industry and infrastructure already exist in mature form. So, compared to the situation in the Unites States, engineering may be seen as a profession of very high prestige. Development needs engineering, and relatively speaking, developing economies may offer more and better opportunities for engineers than for attorneys, accountants, hedge-fund managers, public relations specialists and the like. People coming from better-off families may be more likely to see themselves as business owners, so be more likely to see engineering as preparation to start a business rather than are their peers from middle-class families. It's established that foreign engineers are more likely to start businesses than those native-born. According to the Population Research Bureau, "Foreign-born entrepreneurs helped start one-fourth of all new U.S. engineering and technology businesses established between 1995 and 2005, including Google and eBay."[15] (But note that "helped start" gives credit to foreign-born engineers for startups that also included US-born co-founders.)

Foreign students also know that due to the large number of young engineers that the United States takes in, their chances of emigration to the US are improved if they study engineering rather than, say, public relations.

Foreign-born and-educated engineers don't always bear the burdens of loan repayment that many American college graduates, whether engineers or not, have to carry. They can come to the United States as graduate students, receive support during their studies by teaching or participating in research, and then graduate with advanced degrees with little or no debt. Without the need to service large loans, they are thus better positioned to survive financially through times of low income than are their American-born peers. This may lead to a greater   tolerance for risk among foreign-born engineers.

Employers would have us believe that the reason for the disproportionate representation of engineers born outside of the United States is that they are smarter than their US counterparts, work harder, or both. Reports of frightening deficiencies in the quality of the US educational system   appear   with   some   frequency,   but   have been   strongly challenged.[16]

Is engineering a profession?

If you were to ask a plumber what her profession is, the answer you would expect is "plumber". We understand that when an exasperated homeowner is unsuccessful in finding the cause of a leak and exclaims "we need to call in a professional" she is not talking about her dentist. Yet plumbing is a skilled trade, certainly one for which considerable training and a license are required, but not a true profession in the sense that law and medicine are.

Some non-English-speaking countries use the "Ing." prefix (e.g. Ingeniero in Spanish, Ingenieur in German), or some variant like Dipl.

Ing. (e.g. Diplôme d'Ingénieur in French) as a prefix or suffix, to indicate professional status as an engineer. National rules vary, but for the most part these credentials require more than a four-year degree.

There has been a long-standing debate among engineers about whether we are indeed true professionals as physicians and lawyers are, or simply workers in occupations. Engineers take professional responsibility (but not always legal responsibility) for the design and performance of machines, systems and processes, whose failure to perform properly can expose workers, organizations and the public to immense economic loss, as well as the risk of injuries, sometimes mass injuries. The engineer's failure to properly perform his duty can result in economic and human damage; imagine a chemical plant exploding as a result of a design error, or an engineer's failure to exercise proper supervision over its construction. Engineers working under professional licenses have lost them after their designs' structural failures. The damage can be well in excess of that which can be attributed to a careless medical doctor, for example, who (except for obstetricians and transplant surgeons, possibly) would find it difficult to injure more than one person at a time.

Bus drivers are responsible for the safety of their passengers and others whom they might encounter while driving. They can be – and have been – responsible for instances of multiple deaths, but even though they require state licenses, their vocation is not considered a profession. The phrase "professional bus driver" may not seem odd to the reader; it means simply that the individual drives the bus as an occupation, even if part-time, and not as an occasional activity. Similar is the distinction between "professional musician" and "amateur musician".

The term "learned profession" is occasionally applied to note the distinction between, say, bus drivers and lawyers. This is the sense in which the term "professional" is used in U.S. law:[17]

To qualify for the learned professional exemption, an employee's primary duty must be the performance of work requiring advanced knowledge in a field of science or learning customarily acquired by a prolonged course of specialized intellectual instruction. ...

The phrase "field of science or learning" includes the traditional professions of law, medicine, theology, accounting, actuarial computation, engineering, architecture, teaching, various types of physical, chemical and biological sciences, pharmacy and other similar occupations that have a recognized professional status as distinguished from the mechanical arts or skilled trades where in some instances the knowledge is of a fairly advanced type, but is not in a field of science or learning.

Thus by legal definition US engineers are professionals.

Not quite, however. The Federal law reprinted above is very limited in its application. It was used in the definition of exceptions to the Federal Wage and Hour Law, and in so doing established that employers didn't have to pay engineers (since they are "professionals") for overtime. This is the origin of the distinction between the "non-exempt" technician and the "exempt" engineer, terms that many will be familiar with. Outside sales representatives wouldn't ordinarily be considered learned professionals, and are equally exempt. There is nothing in the law that guarantees or even encourages professional recognition to engineers.

This law has an interesting reference to software engineering: "Computer systems analysts, computer programmers, software engineers or other similarly skilled workers in the computer field are eligible for exemption as professionals...".[18] So at least for purposes of wage and hour laws, which establish that software engineers need not be paid overtime, software engineering is a profession as much as traditional engineering. It's not clear, though, that it's the same profession.

This is a nation where higher education has become rather common. There is a body of thought that holds that to be a professional, one must have completed not only a four-year college program, but a post-graduate program in a "professional school" as well. This is true for medicine, dentistry and law, and increasingly so for other disciplines such as pharmacy and accounting. While many engineers do go on to graduate school, a graduate degree is not currently required for a Professional Engineer license nor is it generally required for an entry-level engineering job as it would be for an entry-level job as an attorney (internships excluded, of course). Thus, by these criteria, an engineer would not be a professional whereas a librarian (with a Master of Library Science as the necessary credential for the field), social worker (Master of Social Work), and teacher (Master of Arts in Teaching, or Master of Education) all would be.

Consider the title "Professional Engineer", which is controlled by state licensing boards and should remove all ambiguity about the professional status of engineers. To become a Professional Engineer one must generally have at least a four-year degree from an accredited engineering school, pass two days of difficult tests, and have several years of experience in the field. This is a high bar. (As noted, state regulations often allow for some flexibility, such as by allowing licensure of holders of certain other degrees once they accumulate additional experience. Equivalent alternative paths are rarely if ever available for prospective physicians, dentists and lawyers.) But the nature of the title begs the question of who are the nonprofessional engineers. Our society has attorneys *licensed to practice law*, doctors *licensed to practice medicine*, *certified public* (not "professional") *accountants*, *ordained clergy* distinct from lay leaders, and so on. It is presumed in the language that doctors and lawyers engaged in practice (as opposed to, say, working in support roles until their bar exam results come back) are professionals, therefore the designation "professional attorney" would be a tautology. Of course the lawyer is a professional. There is no such presumption in engineering, so the term "Professional Engineer" as defined and regulated under the laws of each state, distinguishes the individuals so licensed from other engineers, who are not recognized as professionals. But are these other engineers then amateurs?

Licensing of engineers was first proposed by Wyoming's State Engineer, Clarence T. Johnson, who found that maps for proposed irrigation projects were often of low quality, having been prepared by unqualified individuals.[19] The first Professional Engineer license was then issued in that state to Charles Bellamy on August 8, 1907. By that year there were certainly trained engineers in the United States, with the first degree in civil engineering having been granted almost 75 years earlier, by what is now the Renssalaer Polytechnic Institute.[20] Degrees in other branches of engineering were granted by a number of institutions before 1900 as well. But before 1907 there were no legal restrictions against the practice of engineering by unqualified individuals.

The first Wyoming law for licensing engineers was narrow in its scope. Next came Louisiana, whose registration law was broader. It took many decades for the licensure of engineers to expand in scope to cover other branches of engineering, and to expand geographically to cover all of the United States. The last jurisdiction to enact a law requiring the licensing of engineers was the District of Columbia, in 1950. Even though there is a growing sense that too many occupations require excessive levels of preparation (in one state, 1000 hours of training and two years of experience are needed to become a licensed cosmetologist) the licensing regime for engineers appears secure.

A person qualifying for a Professional Engineering license in one state (or equivalent jurisdiction such as the District of Columbia) might not be able to qualify in another, since each such jurisdiction has the exclusive power to regulate its own engineering profession. States are free to impose any standards they like, including requirements unrelated to the practice of engineering; at least one state won't renew a license for an individual who is delinquent in child support payments.

To maintain some consistency in their requirements, licensing boards generally work together through the National Council of Examiners for Engineering and Surveying (NCEES), but they are always free to deviate from the NCEES models as they wish. For example, state boards may choose to impose, or not impose, requirements for continuing education as a condition of license renewal. While states frequently offer comity (meaning courtesy or respect) to each other's licensed engineers, one still has to obtain a license from each state in which he wishes to practice. True reciprocity is what is extended to automobile drivers; with a license from my home state I can drive in yours without getting a second license, as you can in mine. Under comity you will have to fill out forms, get references, and of course send money, but happily you should not need to retake the engineering exams. Licensed engineers also have the option of applying for "Model Law Engineer" (MLE) designation through the NCEES which may simplify and expedite but not obviate the process of extending licensure to multiple states.

State jurisdiction does not necessarily extend to engineers working on federal projects, although some federal agencies choose to require licensure for specified positions and functions. (Federal positions that advertise for "professional engineers", not capitalized, may in some cases be open to unlicensed personnel.)[21]

With linguistic ambiguity in both "engineer" and "professional", the American English language gives us few tools for distinguishing those whom we would recognize as engineers from those we wouldn't. Not every engineer has to be a licensed Professional Engineer and in fact many are not. The difference between "professional engineer" and "Professional Engineer" is too subtle to be of any use, and use of even the uncapitalized version of the term could subject an unlicensed individual to legal complaint. An "engineering professional" isn't necessarily an engineer at all, just as an "allied medical professional" isn't a physician. A "degreed engineer" might possibly have a degree in another field. "Graduate engineer" could work, but may be awkward in conversation, and can be interpreted to imply that the individual so titled is a graduate student in engineering or has a graduate degree. Perhaps the best we can do is state our branch, as "I'm a chemical engineer". This seems to be the identification of choice for many engineers, but primary identification with a branch of engineering rather than engineering as a whole, carries its own problems, as will be explained later.

My view is that the term "licensed engineer" would be preferable to "professional engineer", in that it references the privileges and responsibilities that are associated with licensure without implying that engineers without licenses can't perform in a professional manner. There is some precedent for this in the title held by some social workers, Licensed Independent Clinical Social Worker, LICSW. I would then use a term like "board certified", borrowed from medicine, or "professionally certified" for those who have demonstrated specialized qualifications beyond the requirements for a basic license. Such certifications could be conferred upon licensed engineers (only) by educational institutions and professional societies, not necessarily by governments.

Most engineers go through their entire careers without professional licensure, and it's all proper. This is because of the "industrial exemption" in most state licensing laws. New York's, for instance, exempts from Professional Engineering licensure requirements "[t]he practice of engineering by a manufacturing corporation or by employees of such corporation, or use of the title 'engineer' by such employees, in connection with or incidental to goods produced by, or sold by, or nonengineering services rendered by, such corporation or its manufacturing affiliates".[22] In practice the exemption is often applied broadly so that, for example, an application engineer employed by a product's manufacturer to help customers install, configure, use and/or maintain the product would not need to be licensed.

A corporation whose business exempts it from the need for licensure is free to give the title "engineer" to whomever it chooses. Engineers most in need of, and therefore most interested in licensure, tend to be civil engineers as well as those in other disciplines whose work involves construction. For example, an electrical engineer designing the lighting system for a building would likely need to be licensed, but an electrical engineer employed to design a lighting fixture to be specified by the licensed engineer probably wouldn't need a license. Even a state that has no industrial exemption is unlikely to check into the professional credentials of the designer of the lighting product. (In practice, independent product certifications such as by Underwriters Laboratories serve a similar purpose.)

Product design and manufacturing processes are complex and usually the responsibility of teams, under corporate control. It would often be difficult to assign responsibility for a product failure to one person individually other than an executive. If a manufacturer, for example, decides to save money by skimping on testing, the product's design engineer should not be required by his conditions of licensure to do the testing on his own time. Also, it's the manufacturer that sets the specifications for the product that the engineer is to design. If an appliance manufacturer decides that its business will benefit from shortening the lifespan of its products,[23] it would be difficult to ethically obligate the engineer designing the appliance to use more expensive parts, without authorization, to make the products last longer. Therefore it's the manufacturer who has the liability exposure for defective products. As always, there may be exceptions and laws that vary among jurisdictions. If this becomes important to you then you need to consult a lawyer.

The industrial exemption, while I understand it, acts in my opinion to disadvantage engineers. Primarily, it has allowed manufacturers, who these days employ the majority of engineers, to create a subclass of individuals of unverified ability. "Unverified" doesn't mean "low", but it does mean that an opportunity for public protection has been waived. As a practical matter, it also encourages engineering work to be shifted to people, perhaps in other countries, who don't have the qualifications of American Professional Engineers, when employing them is less expensive.

The industrial exemption is common in the US but not in every country. Consider how the Canadian organization Professional Engineers Ontario (PEO) enforces registration under Ontario's Professional Engineers Act:

> Sections 39 and 40 of the Professional Engineers Act give PEO the authority to take enforcement action. Section 40(1) sets maximum fines for individuals and firms practising without the necessary licences (P.Eng., and/or C of A). These fines are $25,000 for a first offence and $50,000 for each subsequent offence. Section 40(2) authorizes fines of up to $10,000 for a first offence and $25,000 for each subsequent offence, for people who:

> - use the title "professional engineer" or an abbreviation or variation as an occupational designation;

> - use a term, title or description that will lead to the belief that the person may engage in the practice of professional engineering;

> - ...

> When a potential enforcement matter is brought to PEO's attention, staff investigates and takes appropriate action. PEO also takes proactive measures to address unlicensed activity. In the past this included checking listings in Yellow Page directories, and classified ads in major daily newspapers, to ensure that the use of "engineer" and "engineering" was compliant with the requirements under the Act.

> With the advent of corporate websites, job posting websites and social media, proactive measures have

been adapted accordingly:..."[24]

In Canada, education and licensure are more closely integrated than they are here; when you get a degree from an accredited engineering school in Ontario you won't need have your knowledge of engineering tested again to qualify for a license. You will, though, have to pass a national test in ethics and related material.

Other exemptions to requirements for licensure are controlled by individual states and are not uniform. Some states may, for example, grant exemptions to employees of utility companies and public agencies who would otherwise need to be licensed. There appear to be no similar exemptions for other professions; for example, my research has found no instance of an industrial exemption for architects. Such an exemption, if it existed, might apply to an individual designing an office building intended to be occupied only by employees of the same manufacturing company. Similarly, it is unlikely that a manufacturer through an in-house health care plan would be allowed to employ an unlicensed surgeon under the stipulation that surgery would be performed only within the manufacturer's facilities and on the manufacturer's employees. One reason sometimes cited for the industrial exemption is that the engineer's employer, not a state licensing board, is best able to judge his qualifications. However, society does not yield similar powers to the managing partner of a law firm or the chief of a department in a hospital; the subordinate lawyers and doctors still need government-granted licenses.

Exemptions to licensing laws are likely the result of lobbying by organizations that believed it would be an unnecessary cost to pay a premium to licensed engineers. This can be a false economy. On September 13, 2018, a gas distribution system in Massachusetts became over-pressurized, causing some connected appliances to fail. These failures resulted in a number of fires, one death, other injuries, and property damage that included the destruction of at least five homes. Gas service to the affected communities was lost for months. Many residents were left without heat or hot water and forced to move to temporary housing.

The National Transportation Safety Board (NTSB) found that the overpressure condition arose from work being performed on the distribution system. A section of pipe containing pressure-sensing lines was abandoned and allowed to reach atmospheric pressure. Since these lines now reported low pressure, automatic regulators used to control line pressure acted to admit more gas. This process continued until the regulators opened fully, allowing upstream pressure to reach the distribution system. The work plan had been reviewed by an unlicensed field engineer who did not realize that the construction process would cause the pressure control to react in the way it did. The NTSB's advice to the Commonwealth of Massachusetts was short and explicit: "Eliminate the professional engineer licensure exemption for public utility work and require a professional engineer's seal on public utility engineering drawings."[25] Massachusetts did subsequently change its laws.

My conclusion is that, notwithstanding the privilege that some engineers have to use the title "Professional", engineering in the United States unfortunately does not function as a profession. Entry is not effectively controlled because of the massive manufacturing exemption, with the result that there are no clear qualifications required. Even the qualifications for Professional Engineer licensure vary among the states. Unlicensed engineers have no clear duty to serve the public first; instead they might be presumed to give precedence to their employers' interests. That's not to say that unlicensed individuals are any more inclined to act unethically than those with licenses; it's just that the ethical duties of the former are not so clearly defined.

As an employed engineer, do you want to know whether you're working in a professional role? Ask yourself what would happen if you were to refuse, for good reason, to approve something that your management wants. An auditor who for good reason declines to endorse a financial statement might lose a client but will probably be working as an auditor the next day. The lawyer who refuses a client request because it would require that she violate legal ethics will probably still be working as a lawyer the next day. What about you?

Law professor Paul M. Spinden, in a lengthy review of the history of licensing and the industrial exemption in particular, concludes starkly that the industrial exemption has obviated any claim that engineering has on being a true profession.[26]

Professional Engineers who act as such, e.g., apply their seals on a drawing to indicate approval, must be aware of any liability exposure that such an act engenders. It may be necessary for them to carry insurance or otherwise act to limit liability. Talk to your lawyer.

Do professional status and licensure matter?

From time to time in this text I refer to engineering as a profession, but only as a stylistic convention, with a meaning closer to "vocation" or "occupation" than to a suggestion that engineers, especially those without licenses, are members of a recognized profession.

The licensure of only a minority of US engineers (generally estimated at 20 – 30 percent overall, with civil engineers predominating) likely prevents the more widespread acceptance of engineering as a profession.

In my opinion, engineers' lack of true professional standing disadvantages American society as a whole. If the majority of engineers were recognized and organized as professionals, engineers' employers would be forced to deal with an authoritative, independent body with influence over professional standards and practices. Most professional codes of ethics prescribe a primary duty to the public, rather than to an employer. The National Society of Professional Engineers' Code of Ethics requires the Professional Engineer to "[h]old paramount the safety, health and welfare of the public".[27] Mandatory adherence to such a Code of Ethics would give the engineer additional authority to oppose his management's directives when in the engineer's judgment complying with them could lead to unnecessary risk. True professional status might help protect the engineer from being terminated for insubordination – and protect the public from a bad product. This is the case even though an engineer may not be specifically trained to evaluate environmental and safety trade-offs, for example, to balance the environmental degradation a new electric transmission line would bring, with the improved public safety that would result from a more reliable power supply. We still listen respectfully to doctors on public policy issues, even though members of that profession have arguably overused antibiotics, helping to create new public health problems by the evolution of resistant strains of bacteria. And the tragic consequences of excessive prescriptions of opioids, in some cases by practitioners who weren't told about them, didn't understand them, or perhaps weren't concerned with the addictive nature of these drugs, are unfortunately well known. There are a number of additional consequences of engineers' uncertain standing in society:

We are not taught by our fellows. One expects that the people who teach law students have for the most part practiced some law (and perhaps still do), and that clinical professors of medicine are in fact practicing physicians. Accounting professors are usually accountants, and prospective clergy are normally trained by ordained religious leaders. But the professors of engineering who teach us have often never seriously done what they purport to be training their students to do. The critical credential is the doctoral degree, not the PE license. Standards for program accreditation do not require that faculty members be licensed engineers or have any experience in industry.

It is possible to go through an accredited four-year undergraduate engineering program without once hearing about the possibility of professional licensure. I know that because licensure never came up in my own undergraduate education. Engineering professors on the whole, at least the ones I met as an undergraduate, are more interested in research projects than in engineering practice, a situation that does not engender objections from their university employers since research encourages research grants which in turn lead to publications and prizes. For this reason, among others, engineering curricula tend to be "light" in such areas as engineering ethics, presentation skills, project management and career management, and sometimes professional licensure, all subjects that could benefit both the prospective engineer and American society as a whole. The National Society of Professional Engineers "encourages" legal requirements that professors teaching upper-level courses (as well as engineering school administrators) have licenses,[28] but of course it has no power to enforce this. Individual states may impose such requirements if they choose, but if some states require faculty licensure and others don't, it can become difficult for an engineer educated in a state without such a requirement, to qualify for licensure in a state that does.

Our fellows do not identify with the rest of us. During the early stages of this book's preparation there were two very prominent physicians in American national politics: Bill Frist, a Republican, Senate Majority Leader; and Howard Dean, candidate for the Democratic presidential nomination and then Chairman of the Democratic National Committee. (Yes, this book was a long time coming.) It was well known to those who followed politics that both are physicians; Dean's official candidate biography identified him as "Howard Dean, M.D.". There are certainly engineers who have reached leadership positions in American life as well, but for the most part they do not bother to identify themselves as such. Sometimes they will refer to their backgrounds, whether they ever practiced or not, to score debating points, as in "As an engineer, I understand how to interpret data and you do not". I knew at least one project manager who, to facilitate his climb up the management ladder at one of America's largest corporations (and a technology leader at that), chose to keep the designation "PE" off of his business card. He did not wish to highlight the fact that he was once a competent, practicing engineer out of concern that he would then look like less of a manager. When engineers in general leadership positions do refer to their technical backgrounds, it's often as "former" engineers, whereas the distinguished Drs. Frist and Dean did not identify themselves as former physicians.

We do not regulate our own professional lives. The Professional Engineers who sit on state licensing boards have nothing to say about the activities of the majority of the country's engineers, because the majority of the country's engineers are not answerable to their licensing boards. More generally, engineers do not set the standards for engineering. (This is a reference to technical standards, such as for WiFi, rather than licensure.) In my experience as a one-time standards writer, engineers who sit on standard-setting committees are presumed to carry a duty of loyalty to their employers, rather than to the engineering profession or to the public. Conversely, recognized professionals such as attorneys who work on practice standards are often presumed to be acting on behalf of their professions and the public at large, and not their law firms, individual clients, or other employers even though of course they may reflect their clients' perspectives. They are often identified as, for instance, "a patent attorney" or "an attorney with..." either of which encourages the perception that they are fundamentally independent professionals, rather than employees, to whom the term "servant" is frequently applied.

We do not regulate our own education. The power to confer accreditation to degree-granting programs flows from government, specifically the U.S. Department of Education. That department grants to a number of organizations the privilege of accrediting specialized educational programs, including law and medicine, as well as business, nursing, teaching and so on.[29] Practicing lawyers can exert influence over accreditation through the American Bar Association (ABA). While many members of the Section of Legal Education and Admissions of the ABA are academics, others are judges and practicing attorneys.[30]

There is a similar group for accrediting engineering schools – the Accreditation Board for Engineering and Technology, ABET. The ABET Engineering Accreditation Commission (EAC) is responsible for accrediting engineering schools.[31] When I checked, the EAC Executive Committee had 23 members, of whom 18 had educational affiliations, one had a clear industry affiliation, one represented a professional organization, one was a "public commissioner", and the affiliations of two weren't clear. Practicing engineers have only a minor role in the way that engineering programs are accredited.

Our ability to influence public policy is limited. Engineering societies may argue that they do try to represent engineers as a whole, for example by testifying before Congress and communicating with state legislators on issues of interest to the field. But professionals – like lawyers, physicians, psychologists and (retired) military officers – often act as "talking heads" in the general media, under the presumption that they are speaking not only for themselves but to an extent as representatives of their profession, one that that possesses a body of specialized knowledge.

Engineers are not well represented in regulatory bodies. The Federal Communications Commission (FCC) and Food and Drug Administration (FDA) have responsibilities that, at a high level, are generally similar. Both are responsible for protecting the public by mediating conflicts between commercial interests and the public good, and both preside over areas of commerce that involve specialized technology. As of January of 2021, here are the five FCC Commissioners, their previous occupations, and their post-secondary institutions:[32]

> Ajit Pai (Chairman): Attorney, Harvard University and University of Chicago Law School

> Nathan Simington (Commissioner): Attorney, University of Rochester, Lawrence University and University of Michigan Law School

> Brendan Carr (Commissioner): Attorney, Georgetown University, Catholic University of America

Jessica Rosenworcil (Commissioner): Attorney,
Wesleyan University, New York University School of
Law

Geoffry Starks (Commissioner): Attorney, Harvard
College, Yale Law School

Of the five commissioners, all are lawyers. While their undergraduate
majors are not stated, there is no indication in any of their official bios
that any one of the commissioners has any education or experience in
engineering.

The FDA is organized differently from the FCC, with one Commissioner, one Principal Deputy Commissioner, and four Deputy Commissioners:

Stephen M. Hahn, M.D. (Commissioner): Oncologist,
Rice University (Biology), Temple University (Lewis
Katz School of Medicine)

Amy Abernathy, M.D., Ph.D (Principal Deputy
Commissioner): Hematologist/oncologist, University of
Pennsylvania, Flinders University (Australia, Ph.D.),
Duke University (MD)

Anna Abram (Deputy Commissioner for Policy,
Legislation and International Affairs): Policy specialist,
University of Pennsylvania

Anand Shah, M.D. (Deputy Commissioner for Medical
and Scientific Affairs): Radiation oncologist, Duke
University, University of Pennsylvania (M.D.), Harvard
School of Public Health (Master of Public Health,
M.P.H.)

> Frank Yiannis (Deputy Commissioner for Food Policy
> and Response): Microbiologist, University of Central
> Florida (Microbiology), University of South Florida
> (M.P.H.)

> James Sigg (Deputy Commissioner for Operations and
> Chief Operating Officer): B.S. in Business Management,
> Saint Francis College.

Four of the seven top leaders of the FDA are physicians or scientists, trained in subject matter applicable to the industries they regulate.

Similar patterns are found in other federal regulatory agencies: The National Institute of Standards and Technology (formerly the National Bureau of Standards, note "Technology" is now part of the name) is headed by a Director who is a chemist, assisted by three Associate Directors, respectively a physicist, a patent attorney with an undergraduate major in chemistry, and an MBA with undergraduate training in chemistry. The Environmental Protection Agency is headed by a lawyer who studied English and biology as an undergraduate. The Nuclear Regulatory Commission is a little better; among its top five leaders are two nuclear engineers. The Federal Highway Administration was headed by a lawyer[*] (the Administrator) and a political science major (Deputy Administrator), although the Executive Director is a civil engineer.

Early in the administration of Donald Trump, he was widely criticized for selecting people without science backgrounds for senior positions that traditionally required them.[33] As of this writing it is early in the Biden administration, but Heidi Shyu, trained as an engineer, was confirmed as Undersecretary of Defense for Research and Engineering.

---

[*] The FHWA Administrator resigned in early January 2021

Technology is an important part of contemporary American society, yet it's rare that we see an engineer speaking as an expert on some issue of public policy. Media outlets often have legal, medical and even entertainment correspondents, but notwithstanding the importance of engineering to America, if they have "technology correspondents" at all they're likely to be concerned primarily with the generation and delivery of content for entertainment, or issues of computer hacking. Some of this can be explained in terms of engineering not being a recognized profession, since without a profession to represent, the engineer would be speaking only for himself, and that diminishes the media's interest in what he has to say.

There are discussions that American society should be engaging in, and could be engaging in if America's engineers had a voice independent of our employers. As an example, American appliance manufacturers have been criticized in recent years for cheapening the build quality of their products. Just as doctors can comment on medical protocols and lawyers can comment on the effects of proposed laws on society, engineers should be able to engage in a public discussion – not just within the confines of engineering publications, blogs and so on – on the ethics and appropriateness of designing a product to fail just after its warranty expires, or of designing it to be so expensive to service that the owner has little choice but to replace it when it develops a problem, or of offering small, simple parts for sale only as components of expensive assemblies. I'm not going to take a position here on whether engineers as a profession should try to draw a line, or where such a line might then be drawn. I merely suggest that an independent profession of engineering should, through its elected leadership and peer-recognized experts, be able to discuss such matters with the public without fear of retribution from employers and others.

We may not be considered professionals by the military. It goes without saying that the US military depends heavily on technology and those who understand technology. However while direct commissions are offered routinely to doctors, lawyers and clergy, engineers are more likely to have to go through the traditional officer training process.[34] Policies vary between services and may change over time, so if you're interested in a career in the military you should directly contact the branch or branches you're considering.

We commonly report to people who aren't engineers. Lawyers and auditors often create their own firms, managed by members of the same profession all the way to the top. Physicians similarly are often members of self-managed practices and some become hospital administrators. Engineers do this as well, of course, but in corporate America, working engineers typically report to engineering managers, who then often eventually report to nontechnical executives.

We have trouble identifying and pursuing our own interests. We engineers are joiners, to be sure, but our "professional societies" have at times left a lot to be desired in the way they pursue the professional and economic interests of their members.

We have little political clout. Other professions have done better at having their concerns written into laws and public policy. The American Medical Association and its subsidiaries spend about $20 million every year lobbying the federal government.[35] IEEE-USA, with by my estimate approximately 200,000 members, spent $20,000 on lobbying in 2018 and as of this writing appears to have spent nothing since.[36] However, this simple comparison does not represent the entirety of our profession's political activities. For example some engineering organizations operate political action committees (PACs) and some encourage individual members to communicate with their representatives in Washington and state capitals.)

Our work is not understood by the public. I'm old enough to remember early television accounts of the U.S. space program that predominantly referred to scientists as the people who designed the program's hardware, and technicians as the people who made hands-on adjustments and repairs. Media treatment of engineers is somewhat better now, but still has a long way to go. Sometimes the word "engineered" is used in a very negative way that has nothing to do with the engineering profession, such as the occasional news report that someone "engineered" a cover-up or a corporate coup. I fear that this can affect the public's perception of engineering as a whole.

Our work and experience are undervalued. Much of engineering has become commoditized. Engineering services are fungible – that is, within a specialty one hour of engineering service is often considered roughly equal to another – and there is little credit given for knowledge and experience beyond the minimum. Imagine that you need to hire a plumber. You might choose a plumber based on your previous experience with her, a neighbor's recommendation, price, or the time it takes for the plumber to return your call. You're generally not too concerned about whether the plumber can do the job, since the plumber's license gives you that assurance. You may be concerned about cost but you're likely to find that all licensed plumbers within your market area charge generally similar rates. So one licensed plumber is in many ways equivalent to any other, and you're free to base your decision on secondary criteria.

Now, imagine that instead of a plumber you need a lawyer. In this case the mere possession of a license to practice law may not be enough; you're likely to be much more careful about the lawyer's qualifications and experience.

Your perception of the relationship between price and value is very different in the two cases. When you hire the plumber, your basis for decision might be phrased as "I want to hire the least expensive plumber since I believe any licensed plumber can do the job well." In the case of the lawyer, the basis for decision might be "I want to hire the best lawyer I can afford!" Engineers are largely hired and compensated in accordance with the first model: Human Resources departments may remove Masters Degree requirements to create a larger pool of prospective employees, and to avoid paying a premium for the advanced degree. There's also no reason to pay a premium for an extra twenty years' experience, if the job doesn't seem to need it. Employers often assume that an engineer's productivity peaks after a decade or two of practice, and are therefore reluctant to grant more than nominal salary increases. This, of course, is the philosophical underpinning of the "overqualified" label so often given to engineers, but applied less frequently to attorneys, physicians, professors and so on.

We can't set our own rates. This is part of the previous discussion. Not every accountant works for an accounting firm, just as not every lawyer works for a law firm and not every doctor is part of a self-managed medical practice. But enough do, so that these professional corporations influence the working environment of peers employed by other kinds of organizations. A corporation that wishes to hire a capable General Counsel will be competing for that talent with law firms that the lawyers themselves control, so salaries are likely to be set higher than if corporations dominated hiring. Conversely, while there are certainly engineering firms, engineering compensation is dominated by employers that engineers don't control, which tends to push salaries down. This is only made worse by the irrelevance of engineering licenses for most engineering jobs, a situation not faced by attorneys, MD's and CPA's.

There are weak legal guidelines establishing who may and who may not practice engineering. Our society may celebrate self-educated engineers, but is likely to prosecute self-educated physicians and attorneys. More to the point, weak licensing requirements make it easy for corporations to lower their cost of engineering work by bringing foreign engineers into our economy and sending engineering work offshore. Lawyers and doctors, among others, subject to licensing regimes that are more strictly enforced, don't have to be so concerned about this form of competition. Anyone, citizen or immigrant, holding a foreign medical degree needs to meet American qualifications before being allowed to practice here. But the doctors' turn has now come as well, thanks to medical tourism. There have also been reports of X-rays being sent from US hospitals to be read by US-licensed doctors overseas. Increasing use of telemedicine, replacing in-person visits with technology-mediated remote consultations, may further enable the offshoring of medical services, again subject to state licensing requirements. As of this writing, it does not appear that these phenomena are affecting members of other professions as much as engineers.

We tend to be "trained" more than "educated", and career decisions have to be made early. Doctors, dentists, lawyers, MBA's and other professionals in the United States train for their vocations after graduating from college. Their post-graduate interests may drive their choices of undergraduate courses, but for the most part they are free to select any major. Decisions about postgraduate training can be deferred until the student is well along in her college career. For engineers, it's possible at some schools to enroll in a general undergraduate program with an engineering major, but for the most part an accredited undergraduate education in engineering necessitates enrollment in either a College of Engineering within a university, or a separate institution with a name like "Institute of Technology" or "Polytechnic Institute". Thus in many cases young people are pushed to make career decisions while still in high school. This may be one of the reasons for the notoriously high dropout rates from many engineering programs; it's not just the difficulty of the work, but the fact that young students may not have the knowledge and the maturity to make the best decisions at the time they have to make them.

Professional Ethics

Professions, acting through their professional organizations, create codes of ethical conduct to which their members are obligated to adhere. Ethical codes in the United States generally are voluntary and don't have the force of law in that a violator would risk prosecution, but violation can lead to a loss of license to practice. (A violator might however be subject to prosecution under laws such as addressing negligence.) The National Society of Professional Engineers (NSPE) has a code of ethics, as already discussed, and engineering societies devoted to specific branches often have their own codes as well. Public employees and contractors working on government business may have ethical obligations that are legally enforceable.

A primary canon of engineering ethical codes is an explicit duty to the public which supersedes obligations to employers and clients. Other provisions address typical business concerns, such as confidentiality and avoidance of conflicts of interest. Published codes of ethics apply to people operating under licenses as well as those voluntarily committing to codes promulgated by their professional organizations.

Society expects engineers, with or without licenses, to act ethically. This is easy, of course, when ethical behavior is consistent with the employer's short-term interests. When there is a conflict, good judgment should prevail even in the absence of a formal professional code. The official Rogers Commission report on the Space Shuttle Challenger disaster in 1986 records a chilling exchange between two employees of Morton Thiokol, the manufacturer of the shuttle's solid rocket boosters. This exchange took place during a discussion of whether or not the launch should be postponed because of cold temperatures. The engineers were unanimous in recommending a postponement. A Senior Vice President turned to the Vice President and Director of Engineering and told him to "take off his engineering hat and put on his management hat". It is reasonable to infer from this exchange that the Senior VP recognized an ethical duty on the part of his subordinates but elected to use his management authority to override their collective professional judgment. History records that the management decision was to launch, resulting in the catastrophic failure of one of the shuttle's two boosters and the subsequent loss of the crew. The report is a public record available online,[37] and in my opinion its account of the decision to launch should be read by anyone aspiring or claiming to be an engineer.

A more recent case is that of the Boeing 737 MAX. It was grounded worldwide in March of 2019 after two accidents that each killed all on board. The accidents were attributed to the Maneuvering Characteristics Augmentation System (MCAS) which was intended to give the MAX flying characteristics similar to older models of the 737. A Congressional report[38] found plenty of blame to go around: "The MAX crashes were not the result of a singular failure, technical mistake, or mismanaged event. They were the horrific culmination of a series of faulty technical assumptions by Boeing's engineers, a lack of transparency on the part of Boeing's management, and grossly insufficient oversight by the FAA..." As with the Challenger disaster, engineers' concerns were not acted upon: "...During development of the 737 MAX, a Boeing engineer raised safety concerns about MCAS being tied to a single AOA [angle of attack] sensor. Another Boeing engineer raised concerns about not having a synthetic airspeed system on the 737 MAX. Concerns were also raised about the impact of faulty AOA data on MCAS and repetitive MCAS activations on the ability of 737 MAX pilots to maintain control of the aircraft. Ultimately, all of those safety concerns were either inadequately addressed or simply dismissed by Boeing."

While an ethical code constrains the behavior of subscribing professionals, it can also provide authority – "cover," if you like – for the professional to challenge direction from his management when in his judgment such direction is inconsistent with the public interest. Of course, employers have their own codes of conduct as well, and for public employers and government contractors ethical behavior is often enshrined in law. Whether an employer's code of conduct puts the public interest first, or is limited to proscribing behaviors that the employer doesn't like (e.g., doing business with family members, or accepting valuable gifts from suppliers) is up to the employer.

In the case of the Challenger, the engineers appear to have done all they could, given their position in the corporate hierarchy. In that of the 737 MAX the picture is mixed; engineers are criticized for making erroneous assumptions but engineers are also noted to have expressed concerns about about the aircraft's design.

Is an engineer obligated to resign from his job and go public with his complaints if he concludes that his professional judgment has been overridden or ignored? Starting in 2003, some automotive airbag inflators made by the Japanese company Takata exploded on actuation, directing shrapnel toward the driver or passenger. These actuator explosions eventually caused 17 reported deaths in the United States.[39] Some years earlier, Takata had switched away from a synthetic propellant to ammonium nitrate, considerably less expensive but unstable. US chemical engineer Mark Lillie objected to this change for safety reasons, but was unable to convince his management to rescind the decision. He chose not to be associated with a product he considered dangerous, and left Takata. He became a whistleblower, furnishing internal Takata materials to federal prosecutors.[40]

Licensed engineers facing ethical questions may wish to consult with their state licensing boards or a knowledgeable attorney. Another source of guidance applicable in some cases will be professional standards, both voluntary (usually developed by professional societies) and mandatory (like building codes, as well as corporate standards), which play roles similar to that of the "standard of care" in medicine.

I am not aware of any working engineer being expelled from a professional society or criminally prosecuted for making a good-faith error, or for failure to report a problem, but civil lawsuits are not uncommon. Some engineers need to carry professional liability insurance; your attorney will be the authority here. But when an engineer willingly participates in illegal activities, the situation can be very different. In 2016 James Robert Liang, a Volkswagen engineer working at a test facility in California as Leader of Diesel Competency, pleaded guilty to one count of conspiracy to defraud the United States in connection with Volkswagen's efforts to cheat on emissions testing.[41] He was subsequently sentenced to 40 months in federal prison.[42]

Parallel to the situation of unlicensed engineers working in exempt roles is that of licensed engineers working in these roles. Are licensed engineers working in exempt industries subject to professional ethical codes? The NSPE believes so.[43] It recommends that engineers, licensed or not, act in accordance with their employers' ethical guidelines if they exist and are applicable. If required, engineers should first raise any issues with their employers and then as necessary with outside authorities. In extreme cases, the NSPE recommends engaging a private attorney. (My note: Be careful about disclosing your employer's confidential information to outsiders. Your employer might possibly construe such a disclosure as a violation of your confidentiality commitments even if covered by attorney-client privilege.)

Since most engineers don't have explicit ethical obligations, engineering ethics may not be a required subject. A study at the University of Massachusetts published in 2000 found that fewer than 27 percent of engineering schools require every student to take at least one ethics course[44] even though accreditation requires that students acquire "an understanding of professional and ethical responsibility."[45] Few engineering schools seem to have any Professors of Engineering Ethics; ethics courses may instead be taught by law professors. There are also continuing-education courses in ethics available to engineers already in practice.

In serious cases corporations themselves and their senior managers can be held criminally liable. I am not aware of any instance in which a non-management engineer in an exempt industry has been subjected to personal prosecution because of his activity as an engineer, but if you think you could be subject to liability claims or any other legal jeopardy in connection with your work, a consultation with a qualified lawyer would be in order. When an engineer's actions are sufficiently egregious, they become matters for legal resolution. In 2020 in California an engineer pleaded guilty to stealing trade secrets and was given an 18-month prison sentence.

The trickiest situations are probably those that are not clearly addressed by ethical codes or the law but for some reason still give the engineer cause for concern.

Resources on Ethics

The NSPE website (nspe.org) hosts a considerable amount of information on line but not all of it is available to nonmembers. Its library of ethics cases is available to the public but registration is required.

The Online Ethics Library (onlineethics.org), sponsored by the Center for Engineering, Ethics & Society, aggregates links to many sources of information. Also, check providers of free courseware for training material. MIT is one, as of this writing offering a graduate course in engineering ethics through its Open Courseware program.

Resources on Licensure

A good general reference is "A Primer on Engineering Licensure in the United States".[46] For up-to-date information on licensing regulations in your jurisdiction, contact its licensing board. Remember that state laws control the practice of engineering and that there is no requirement that they be identical between jurisdictions. For general information, check the NSPE and NCEES.

Chapter 2: Getting Educated

Having defined, albeit with some ambiguity, who for our purposes is an engineer, let's look at how young people typically prepare for the field. There are actually many options, but I'll concentrate on the usual path of a four-year program of courses.

Preparing for college

Every student is different. You should consult with your guidance councilor, other education professionals and active engineers to help chart your path. That said, the usual advice offered to aspiring engineers is accurate: Take all the math and science you can. If you're not consistently in the top 10 or 20 percent of your high-school class in math and science, and in your other coursework, you might want to consider another career. Conversely, if your high-school guidance counselor tells you "you're good at math and science so you should go to engineering school", don't automatically do so. Maybe you'd be happier in pure science, medicine, finance or some other field where you could make use of many of the same skills.

In math, pay special attention to the kind of problems you see in homework and tests rather than to theory. Learn how to translate the verbal statement of a situation into a problem that you know how to solve. Engineering students often take different math courses than do math and science students. Engineering courses emphasize the applications of math rather than, for example, proving theorems.

Unless you're interested in biomedical engineering, food and pharmaceutical production, or moving on to medicine after college, favor physics and chemistry over biology; but if you have time take them all. Earth science will be useful for those interested in careers in mining or petroleum engineering. It will also be of value to those interested in environmental issues, regardless of what engineering major you may be leaning toward. Environmental science will be valuable for all future engineers, particularly those who expect to work in industries that process materials, that use or produce large amounts of energy or produce significant waste streams.

As for extracurricular activities, there are usually some obvious choices such as competitive robotics. But don't ignore opportunities to pursue other interests, even if unrelated to science, math and technology. Get involved in student-run businesses to gain experience in management. Debate or theater will help you become a better communicator. Learning to play poker could help you become a better negotiator. Or you might just do something because you like it.

Whereas a prospective doctor, lawyer or business manager can reserve her college years for "education" and then her medical, law or business school years for "training", the future engineer must "train" for a responsible job during his four-year baccalaureate program, getting "educated" to the extent possible almost in his spare time. This compression of the engineer's training time has some predictable results. One is that the prospective engineer is forced to make a major career choice early in his education. Traditionally, the second half of an engineer's education is spent specializing in his chosen branch. However, upper-class programs often have specific prerequisites that must be satisfied with courses taken during the first two years. Thus in practice, engineers must often begin to decide on their career directions within engineering–opening some doors but closing others–in their first year in college. In contrast, prospective lawyers, doctors, and managers (MBA's) while generally steering their academic programs toward subjects relevant to their career interests, will have their college years to decide whether to go in one of these directions, and then more years to decide on a specialty, e.g., criminal vs. patent law, or surgery vs. internal medicine. Furthermore, the prospective engineer must in many cases apply to and enroll in a specialized engineering school, whether freestanding (the "institute model") or an engineering school within a university (the "university model"), while others attend liberal arts or arts & sciences programs regardless of whether their interests lie in physics or government. Students in such programs typically don't have to declare their majors until well into their college years. Thus the education system forces engineering students to make their initial career choices as early as age 17 or 18. It shouldn't be surprising then to see a net mass flow out of engineering as students mature and become aware of other opportunities, or simply decide, after initial exposure to engineering, that the field simply isn't for them. While recoverable, this can be a costly error if it extends the time needed in college beyond the nominal four years.

Selecting a college

Engineering schools generally follow either the institute model or university model as discussed above. The first model, exemplified by schools like the Massachusetts Institute of Technology (MIT), defines an institution that is primarily focused on engineering, science and math. The Rensselaer Polytechnic Institute (RPI) and the California Institute of Technology (CalTech) are other well-known schools in this category. Following the second model are schools that include engineering programs within larger universities; examples are Stanford University, the

University of Michigan, and Cornell University. We can further characterize universities by the independence of their engineering programs; some have separate engineering schools while others have engineering departments within their undergraduate colleges. There are overlaps between the models. It is certainly possible at many technical institutes to major in nontechnical subjects. MIT, for example, offers majors in economics, music and several foreign languages, among many others.

You can get an excellent engineering education at a school of either type, but one may be more suitable for you than the other. If you want to take government courses with government majors and psychology with psych majors, for example, a university might better suit your needs. However if you want to learn how arts and literature interact with engineering, an institute could be the better choice. Employers may perceive, rightly or wrongly, that institute-model graduates are better at hands-on engineering, or that university graduates tend to be better communicators and have a broader view of life and society, and are thereby better candidates for promotion to nontechnical management.

Avenues for engineering education can be broken down further than by institute vs. university. The Carnegie Classification of Institutions of Higher Learning[47] maintains a more granular classification system that, for example, includes the category "major research university." Its website includes a useful search facility that, among other things, allows you to quickly identify institutions with characteristics similar to one you name. The site can be helpful for the prospective engineering student who wishes either to apply to a variety of school types or focus the application process on one kind of institution.

Be sure to find out which programs are accredited. I addressed the issue of accreditation earlier, noting that in some cases it's effectively optional and for some courses of study may not even be available. You can go through an entire engineering career without accreditation being an issue, or use your non-accredited engineering education as the basis for a switch to business, law, medicine or some other field. However, a non-accredited degree could become a detriment in a competitive job market, and might complicate the process of becoming licensed. Even if licensure is not required for the career you plan as a high school student, your plan can change. College is expensive and you want to be an informed consumer. I am aware of no discounts for non-accredited degrees. There is at least one program (Dartmouth College) where it's expected that an aspiring engineer will earn a Bachelor of Arts in engineering after four years and then take an additional year to earn a second, accredited degree. A person getting a Bachelor of Arts degree in a related field, such as physics, can also get an accredited degree in engineering after a fifth year.

Large or small? Public or private? Oriented toward teaching or research? Excellent engineering schools are abundant in the United States. Fortunately from the prospective of someone seeking to begin an education in engineering, most states have public universities or technical institutes that are quite solid, and some are among the very top. States often see technology as key to their competitiveness, and may be willing to fund their engineering institutions at a high level despite budget pressures. A non-exhaustive list of outstanding public engineering schools includes the University of Michigan, Georgia Institute of Technology (Georgia Tech), the University of California at Berkeley and Purdue University, the latter not always being recognized as public because its state (Indiana) is not included in its name.

Do you need to go to a top-10 or top-20 engineering school? Of course not; there are differences between programs and the level of competition, but any program with ABET accreditation should be fully credible. A dean of a highly-regarded engineering school told me some years ago that the top engineering schools in China were comparable with the best of the US, but that he considered US schools in the second tier to be far better than their Chinese peers. The quality of professors in all engineering schools should be high since faculty positions are sought after; qualified applicants abound. Employers may, in fact, worry that a graduate of a prestigious program will be more likely to resist management direction or get bored and leave if a more "interesting" opportunity becomes available. And managers who themselves graduated from mid-level programs may be apprehensive of, or even resent subordinates whose degrees are seen as more prestigious. Conversely some employers want the best academic credentials, and a name-brand degree could better help you move into a non-engineering field such as business consulting or investment management. A more prestigious degree could also help you change fields via a professional school.

There is a belief in engineering that "after your first job, it doesn't matter where you went to school." There is even research, not specific to engineering, suggesting that people who are accepted to high-status schools but choose to attend schools that are less prominent end up doing just as well as those who went to the high-status schools. The implication is that the people who become high achievers as adults are also high achievers in high school and even before. Therefore their success has more to do with the prestigious schools recognizing them as such, than their actually having learned more at such schools.[48]

Public institutions in particular often have programs designed to provide employees to local industries but de-emphasize or ignore industries that are not present nearby. Thus at the University of Maine Orono you can find (in the Chemical Engineering department) research programs in papermaking, and at the University of Texas Austin there is a strong interest in petroleum engineering.

You don't have to spend four years in one school to get a degree from that school. Students frequently transfer between institutions for one reason or another. Of particular interest to prospective engineers concerned with cost is the option of getting a two-year associate's degree from a community college and then transferring to a four-year school for your bachelor's degree. You can save a great deal of money at the cost of foregoing two years of campus life. Guaranteed-transfer programs whereby states connect their community colleges and public universities can simplify the application process and reduce the risk of being rejected by the four-year institution. Understand, though, that the guarantee is conditional, typically requiring you to graduate from the two-year school having taken the right set of courses and with a high grade-point average. If you are interested in this kind of program, talk to one or more community colleges, and find out about both their guaranteed programs and any others that may be available. Then visit the four-year campus, seek out some community-college transfer students, and ask what they think of their preparation versus that of peers who started at the four-year school. By doing so, you can determine whether the community college adequately prepares its graduates for the four-year school's upperclass programs.

Starting at a community college will also give you a chance to see what the field is like before committing yourself to it. Technology is everywhere, and many jobs that are traditionally blue-collar also require a good familiarity with the basics of engineering. At a community college you may be able to look in on trade-oriented programs such as for HVAC (heating, ventilation and air conditioning) installation and maintenance to see whether one of these programs might be appealing. Retirements and technology changes are creating many opportunities in these fields, and you may still be able to pursue a four-year degree later.

An emerging option is online education. Educational institutions responded to the COVID-19 pandemic by moving online to the extent they could. Laboratory sessions were replaced by simulations, or by students writing reports of lab work that they remotely observed others doing. As of this writing, the extent to which online options will remain once the pandemic subsides is not clear. Programs in software engineering don't need laboratory facilities beyond a personal computer, so have been transferable more easily than traditional engineering to fully online presentation. Online engineering courses at the undergraduate level in particular may become hybrids, with much of the coursework continuing online but with mandatory on-campus residence or internships. Colleges may become willing to grant credit for completing online courses given by other providers, particularly in liberal-arts electives and non-lab courses like math.

Whether you study at a university or an institute of technology, you'll probably take one non-engineering course per semester if you stay with the traditional educational sequence. Courses in economics and psychology are popular. For some engineering students, choosing a liberal-arts course can be as simple as finding one that fits your schedule and doesn't look like too much work. If your interests are more specific, consider how the college might be able to support them. A university may give you more options than an institute. Don't necessarily limit yourself to the College of Liberal Arts or equivalent; depending on your school's organization and rules, you may be able to meet your outside-of-engineering requirements by taking courses at other specialized schools such as art or journalism. Don't assume your faculty advisor is familiar with every possible option, especially on a large campus.

Whether or not you can do it as an officially-recognized minor, consider putting together a sequence of related non-engineering courses to help open up options later. Obviously, courses in economics can help make you more attractive to a business school, or courses in government and philosophy could make you a better candidate for law school. Foreign-language courses could help prepare you to work with people in other countries, although I've found that engineers worldwide are often able to communicate in English, particularly written English. (Not necessarily so for plant operators, technicians and so forth. It depends on the country and its educational practices.) If you're interested in either work or graduate study in a non-English-speaking country, then of course a knowledge of that country's language will be very helpful. This will be true even if a multinational's official corporate language is English (often the case) or if some of the coursework is available in English.

The college visit

It's been customary for high-school students who can afford to do so, to visit some of the colleges they are considering, whether while deciding where to apply or deciding among schools to which they've been accepted. Physical visits have been curtailed by the coronavirus pandemic, but they will certainly resume. Students unable to visit in person may be able to take advantage of online meetings and/or online participation in classes. Ask guidance counselors for resources to help get maximum value from in-person or online visits. Try to locate some alumni in your area, and if the college provides for interviews with alumni use these as an opportunity for two-way communication.

Here, I'll give you a few tips to help you get the most value from school visits:

Visit the sections. If the college directs you to a large lecture hall packed with people, typically for a class required for first- or second-year students, ask also to visit a section. The professor introduces the class material to all of the course's students at once in a large lecture, but the eminent professor might not be the best teacher, and sections are where questions are answered and homework and test results discussed. See how the section leaders, typically graduate students, interact with their students and how effective they are at getting across difficult points. Section leaders may be foreign graduate students, not native English speakers, who have not been in the United States for very long. See whether they have trouble making themselves understood.

Investigate the school's safety culture. The world is not wholly digital. Through both research and coursework, engineering and science students often work with hazardous materials, machines and processes. A 2010 article in Scientific American[49] recounts a number of serious accidents, including fatal ones, in academic laboratories. In 2011, a Yale student was killed when her hair was caught in a lathe as she worked alone at night, contrary to safety rules.[50] A few years after that I observed, at a well-regarded engineering school, students not wearing safety glasses while they worked on fiberglass. Safety glasses were readily available in the lab but unused. Find out whether safety briefings are required before students can use a lab. What are the rules about students using labs alone, e.g., who can get a key? Are requirements for PPE (Personal Protection Equipment) clearly posted and enforced, and is the required PPE readily available? What are the provisions for containing a release of hazardous materials and assisting those who may have been exposed? Is the lab routinely audited for safety? Are fire extinguishers, first-aid kits and other appropriate equipment and materials readily available, and are students required to know how to use them?

Go outside of the engineering department and try to talk to professors and students elsewhere in the institution if appropriate to your interests.

Find out how other departments treat engineering majors. For example, a popular liberal-arts class with an enrollment maximum may give first preference to students for whom it is required, second preference to non-majors within the liberal arts school, and last preference to students from other schools. Remember also that engineering curricula tend to have a lot of required courses. Other engineering students will likely be taking the same technical courses that you are so there may be a lot of competition for a popular outside course that fits many students' schedules. While the engineering school can reasonably be expected to minimize scheduling conflicts among its own courses, you yourself will have to find non-engineering electives whose schedules are compatible with the engineering courses you're required to take. Larger schools may give you a greater opportunity to choose between different sections of required courses and so give you more flexibility in choosing outside electives.

Learn about the institution's prevailing teaching styles, and determine if they match your style of learning. Certainly different professors teach differently, but there may be an overall pedagogical culture established by deans and department chairs. The usual engineering course comprises the presentation of a set of facts, with homework and perhaps a few worked-out problems, followed by a test which gauges the student's ability to use the presented facts to solve unfamiliar problems. For example, an introductory course in statics might go through the methods of summing forces and moments, and then present the student with situations to analyze; e.g. can this crane hold the specified load or will the load cause it to overturn? For many, this style, called deductive, is perfectly fine and can be the basis of an effective engineering education.

But some of us are motivated by first being given a problem to solve. This is the inductive approach. Some students may pay closer attention to the material when the instructor shows them the tools they will need to solve a problem only *after* they've seen the problem. The inductive approach, combined with project-based learning (students with varying specializations working together to solve a complex problem) can help promote curiosity as well as bridge the demarcations that are often sharply present among academic departments but not always in the real world.

Does the institution make good use of visuals, demonstrations and simulations? Again, undergraduate engineering students are training, in large part, not to be junior scientists but to be practical problem-solvers. Teaching tools that help the equations "come alive" can facilitate learning. Here's a simple rule to help you judge which style is in use: Do demonstrations tend to come after the principles have been thoroughly discussed (deductive) or before (inductive)? Similarly, are lab courses viewed as necessary evils where the student proves that he can follow the procedures presented in class, or is there some opportunity for creativity?

You should have a good idea of your own optimal learning style from your experience in high school, but the raw intelligence you used in high school to "power through" courses taught in the wrong style (for you) could fail you at the college level. It might even be that your high-school teachers used primarily the deductive approach because they knew that not all of their students were highly motivated, but by the time you get to engineering school it's reasonable for your professors to assume uniformly high motivation. This could, in fact, be a contributor the high dropout rates in engineering schools: It's not so much that some students aren't "smart enough" for engineering; it's that the style that feels most natural for engineering professors to teach in may not be the best for all students to learn in. For summary information on how instructors can teach to varying learning styles, see the paper "Learning and Teaching Styles in Engineering Education."[51] For more detail, see the work of Richard Felder, Professor Emeritus at North Carolina State University.[52]

Speak to students who dropped out of engineering. If you're introduced to current engineering students ask whether they have former engineering classmates, still on campus, who switched majors. Find out why they left engineering. Were they not prepared for the program's rigor or demands on their time? Was the school's teaching style incompatible with their learning style? Was it more a matter of their developing an interest in a new major than losing interest in engineering? This line of questioning may lead you to some interesting and valuable information.

Check the alumni magazine. Careful perusal of the class notes in the college's physical or online alumni magazine will give you an idea of the trajectories of its most successful alumni. Look at recent issues, issues from 10 years ago, 20 years ago, and so on. Look at the magazines' feature articles to see what kinds of persons the college considers to be its most noteworthy graduates. Independently of helping you to evaluate the school, this information can help you to plan your career.

Find out how early in your academic career you have to to choose a specialty. An engineering school may allow you to select a specialty starting with your junior year, but that specialty's mandatory prerequisites may force you to make a decision earlier. Even if you think you're certain about the direction you want your engineering career to take, look for a freshman-year course that exposes you to a variety of specialties. You may find that you're interested in something you hadn't thought of or been exposed to before.

Look at the selection of upperclass courses. It's fine if you, as a high-school student physically or virtually touring a campus, don't yet have a sense of what you might want to specialize in after college graduation. Even if you do, your career interests may change as you learn more. But if you do have an idea, based perhaps on a hobby or a summer job, you can look for course options for your junior and senior college years that will help you in that field. Understand, though, that there are no guarantees here: The catalog may list courses that aren't taught every year, the one professor who teaches a course you want could be on sabbatical the year you want to take it, or a course you're interested in could simply be dropped by the time you're ready for it, either for lack of student interest or of someone to teach it.

Find out the fields in which the college focuses its research, and what specifically its research programs are. The obvious reason for doing this is that it will inform you of the professors' interests, which may or may not be aligned with yours, not that new professors couldn't be hired or your interests change. Beyond that, research programs are usually built with grant money from government and industry. That funders invest in a specific subject area implies their expectation that the area will expand and generate revenue and employment opportunities. Find out also whether research programs are narrowly focused or interdisciplinary; I see the former as largely scientific inquiry while the latter are more closely aligned with engineering practice. Also try to learn about research programs that have recently ended; this may indicate diminishing interest on the part of the institution and likely on the part of prospective employers as well.

That said, there are suggestions from research that rapid technological change, while associated with high starting salaries, also leads to salaries quickly leveling off, since accelerated change motivates employers to sideline older employees while energetically recruiting younger ones. I cover this topic in more detail later.

Find out about research opportunities open to undergraduates. If you think you might want to go on to a doctoral program, your application will be stronger if it includes some research experience. Research can also provide you with opportunities for paid work either during the school term or between terms.

Timing is everything. All things being equal, you can learn more by observing classes later in the term. Students, especially freshmen, may feel overwhelmed at the beginning of a term. By observing classes later, when they have reached steady state, you can get a better sense of how students and faculty work together.

Find out how many in the faculty are full-timers versus adjuncts, but don't be upset by the presence of the latter. Colleges are increasingly dependent on adjunct professors – "contingent faculty" – instructors who are paid by the course and typically aren't even full-time employees, let alone tenured or tenure-track professors. The American Association of University Professors (AAUP) estimates that more than half of current academic appointments are for part-time, non-tenure-track positions.[53] In first two years of an engineering curriculum the student gets the fundamentals of science and math which will be the foundation of further study and practice. These courses, in my view, should be led by experienced professional faculty, supported by capable graduate, and sometimes upperclass undergraduate, students. Other underclass courses that are important but not foundational, such as a writing seminar, can reasonably be led by adjuncts and graduate students, although I think that full-time faculty would be preferred here as well.

However, adjuncts can be valuable, and perhaps even preferred, as leaders of some upper-class courses. The reason is that faculty in most institutions are expected to do research as well as teach, whereas adjuncts may be primarily practitioners who work regularly with the same subject matter they teach. Educational institutions have an incentive to hire faculty members who are the most likely to receive research funding from outside agencies. Thus their hiring policies tend to skew faculty interests toward fields that are currently "hot" but more established fields can still present interesting challenges and offer good career prospects. The experts in these more traditional fields may be the people currently in practice outside of academia. Further, the specializations of new doctoral graduates can reflect the priorities of their programs' funders. Faculty composition will evolve as older professors retire and are replaced, and it may be that courses in sub-specialties now in vogue have to be taught by adjuncts if qualified full-time faculty are not yet available. It's also the case that as engineers we can benefit from the perspectives of both academics and practitioners.

What fraction of the engineering faculty are actually licensed to practice engineering, or have practiced in exempt industries? A reasonable number of formerly or currently practicing engineers (the latter often adjuncts), or at least faculty who occasionally go on sabbatical in industry (consulting assignments, in effect), could help balance the academic orientation of your education. Of course, if your plans include advanced graduate education followed by a career in research, then the presence of practitioners may not be so important.

Find out who the school thinks its customers are. Many engineering schools convene advisory panels from industry to help guide curriculum development. It's common for the schools to view educated students as their product and these students' prospective employers as their customers. Of course, there's nothing wrong with the school making an effort to align its educational programs with industry's current needs, and after investing four years and sometimes hundreds of thousands of dollars many new engineers need to move to gainful employment as quickly as they can. But the training we receive should promote our long-term interests as well as our employers' short-term interests, which may be limited to their next products or projects. It can be difficult in a short visit to develop a sense of how the school views its mission, but one thing you can do is find out about its advisory panels. Do they represent a variety of industries, or does one predominate? Are there practicing, non-management engineers, perhaps alumni, on the advisory committees? Ask about how and why the curriculum changed through the past few years.

Investigate the balance between research and practice. This discussion is related to my earlier comments on adjuncts. There isn't a universal right answer, but there may be a right answer for you. Research-oriented programs prepare their students for areas that are developing rapidly, and that we presume will generate a substantial number of jobs in the coming years. But traditional specialties can still provide good careers, and may be more stable since they represent existing businesses and are less dependent on research funding. Chemical Engineering programs are moving toward biotechnology, but we still use a lot of liquid fuels and I'm sure we are going to have oil refineries and bulk plastics plants for years to come. Electrical Engineering curricula have moved toward digital technologies but we still have to generate and distribute electric power. Mechanical Engineering programs may emphasize robotics, but we'll still be using steam power for a while.

Find our how much flexibility you have to change your mind. I've made the case that an accredited degree is the best path for a student who plans to engage in the practice of engineering, but engineering practice isn't for everyone and neither is an accredited degree. You could decide, for example, to forego some required engineering courses in favor of others that might better prepare you for a post-graduate program in another field. What will happen if you decide halfway through to do something else? Will you be able to simply change majors, will you have to apply to a different college within the institution, or will you need to find another school altogether? If you have to apply to another program within the institution, is acceptance guaranteed? How far along in your engineering education will you be able to get without having to spend more time and incur extra expense if you switch to something else? These are answers you should have before you commit to the institution's engineering program.

Where do students come from, and where do they go after graduation? Research has found that students from lower socioeconomic backgrounds tend to remain in engineering after graduation whereas those from society's higher strata are more likely to move on to other fields.[54] Could it be that public institutions, usually less expensive, are more oriented to training working engineers whereas private schools lean toward training students who are more interested in business school, law school, medical school and perhaps academia?

How will the school help you reach your next goal? Of course, all engineering schools will offer you help to arrange internships, find employment after graduation, and gain further education. They do these things through both formal and informal means, the latter including alumni networks. But after visiting a few colleges you may find that some do them more effectively than others.

The school should have a recruiting season during which employer representatives come to campus to interview prospective employees. Local employers should be well represented, but does the school also attract national and global companies? This is not to imply that there's anything wrong with local companies, but a school that also brings in recruiters from far afield can create additional opportunities for you.

Find out about opportunities for relevant employment while you're still in school. Does the school offer a co-op program wherein industry experience is viewed as part of your education rather than as an optional supplement? How many of the companies coming to recruit have openings for underclass or upperclass undergraduates to serve as interns? Give special credit to schools that prohibit unpaid internships by conditioning the opportunity to recruit for post-graduation jobs on the availability of paid internships only.

Is the political environment on campus distracting? Engineering is a difficult course of study. Whether or not you wish to devote your spare time to political activism, do students come under social pressure to do so? Do students frequently drop out to pursue political activities?

How has the school gotten past COVID? I ask this question assuming that it will. I expect that some schools will do better than others in retaining practices that add value while dropping those that reduce value or add cost.

Paying for your education

In this brief section I will talk about some aspects of educational finance from the perspective of a future engineer. There's a lot of general information available on how to finance your higher education. I'm not going to try to duplicate it or dispute it, and whatever I might say probably wouldn't be current by the time you read this anyway.

Middle-income families fall into the "sour spot" of college financing. They are neither rich enough to be able to afford college without a problem, nor poor enough to qualify for massive amounts of financial aid, particularly in the form of grants rather than loans. One attraction of engineering careers for middle-income students is that the field offers a clear career path starting at high school: Take all the science and math courses you can, enroll in and graduate from a four-year engineering program, get a job, and start paying off your loans. In contrast, some other careers require more extensive (and expensive) preparation and/or tend to be reached by paths that are more circuitous or at least less obvious. Just look at the varied backgrounds of the founders of today's tech giants for real-world examples.

Some students take a year or two off after high school or during college to earn money to help pay for their higher education. It is the case, though, that coursework in mathematics and science builds on earlier work, in a way that, for instance, history and English at the undergraduate level probably do not (although foreign language study does). Many have taken time off successfully but it's possible to forget important material before having a chance to use it. Prospective engineering students are likely to have an easier time of it if they go directly from high school to college, without the opportunity afforded by a gap year to work and save some money. If you are thinking of delaying or interrupting your training, ask your prospective or current college about its experience with students in similar situations.

If you go into the military after high school, try to get into a technical specialty where you can get some training and experience relevant to the engineering field you're considering. As a veteran you'll qualify for the standard educational benefits offered by the Department of Veterans Affairs, as well as perhaps an Edith Nourse Rogers STEM scholar-ship[55] which may offer additional support if you study engineering. Check this if you're interested in the military before college.

Borrowing money is one time-honored way to deal with the expense of college. If you reach the limit of what you can borrow from the federal government, there are private loans and parent loans. It has been widely reported that student loan debt in the United States now exceeds credit-card debt. Indeed, it is possible for an American student to accumulate debt in the five-figure range or perhaps into the six figures, before he's old enough to legally drink a glass of beer. Whether this is a good thing or not is open to some debate.

Borrowed money can make it possible for you to attend the school of your dreams. Many years ago I borrowed and I did. But before you go into heavy debt to pay full sticker price at an expensive private institution, I suggest you consider the following:

The ready availability of loan money helps keep educational costs high, or so many analysts believe.[56] Loan money reduces any incentive a college has to reduce costs. This includes administrative costs having only a tenuous relationship to education. Administrative costs are widely reported to be rising more quickly than costs directly related to instruction.[57]

The federal government is making serious money on its college loans. Depending on how you do the accounting, the federal government makes a significant profit on the money it loans students.[58]

Government takes care of its own. Sometimes. Under the Public Service Loan Forgiveness Program[59] government partly forgives loans taken out by people who work for government, as well as certain nonprofits, after graduation. The rules are complex – only the right kind of federal loan qualifies, and you must have already made a significant number of payments before you can take advantage of the program. Participants in the program don't even have to pay federal taxes on the amount forgiven. Unlike many other government programs, there is no income maximum for participation; a high-paying job as an engineer or manager will not disqualify you as long as it's for the right kind of employer.

I have yet to find evidence of a correlation between the expense of an engineering education and subsequent economic success, and in fact there are reasons to believe that excessive debt can impede an engineering career. High levels of debt can be debilitating, and it's my belief that many students underestimate its effects. It's true that many successful companies have been founded by recent college graduates, or indeed by college dropouts. But once you're deeply in debt from engineering school, unless you propose to renege, which I do not in any way advise, you have little choice but to seek the kind of employment that will leave you with a reasonable net income after you make your monthly loan repayment. This typically means, of course, seeking an entry-level engineering job, which will leave you with little time to develop your own ideas, and will likely encumber you with potential claims on your work by your employer. Your dream of creating a company can for many years remain just that if you're burdened by excessive debt. Career flexibility moves inversely with debt.

So engineering students will likely have to borrow significant amounts of money for their educations. It's fortunate, then, that engineering starting salaries tend to be high compared to other fields that employ holders of four-year degrees. As discussed earlier, the people our society recognizes as professionals generally have more than four years of post-secondary education. Many engineers, though, enter the workforce holding no more than a Bachelor's degree.

If you decide, as many have before you, that the rewards of a career in engineering don't justify the requisite cost and effort, you might want to get some more schooling, likely in a "professional" program such as to earn an MBA. The additional debt you incur for this education will be added to the unpaid debt from your undergraduate education to yield an immense sum. This will bother some people more than others. For many, the prospect of, say, doubling personal indebtedness could be enough to push them away from a top-notch professional program toward a more modestly priced one, or may prevent a career change altogether. Of course, you can also work and scrimp for a few years while you pay off your undergraduate loans and then work and scrimp for more years to build up the funds to finance your next round of education.

This is my advice for most prospective engineers: If you can qualify for a brand-name private engineering program and can reasonably afford to attend it without incurring crushing levels of debt, strongly consider doing so. Otherwise, recognize that undergraduate engineering programs, especially those that are accredited, are more alike than they are different. Save your money, or more precisely, examine your willingness to borrow money. You'll be able to use the money for your graduate education, or to support yourself while you work on building the next Microsoft. Good luck, and when you succeed, don't forget who gave you the advice.

Selecting an engineering major

Most engineering schools follow tradition, in that they divide engineering into its four founding branches – Civil, Chemical, Electrical and Mechanical – plus Industrial, and then sometimes combine disciplines from more than one branch to create such fields as Aerospace Engineering, Material Science, and Structural Engineering. Some specialties combine engineering with outside (non-engineering) fields, to create such disciplines as Engineering Economics, Engineering Physics, Biotechnology, Computer Engineering, and Operations Research. Specialized programs are designed to meet the needs of established local industries or help promote nascent industries that political leadership wishes to attract to an area. Examples of the former are Petroleum Engineering, Mining Engineering and Railroad Engineering. Some subject areas will not have accredited programs available just because accreditation criteria have not yet been established.

Students may pick engineering majors as a result of interests developed in high school or perhaps even earlier. This can be good or bad, but is likely to be bad if the school doesn't make a point of exposing its beginning students to the variety of choices that engineering presents. Thus a few words about each of engineering's major disciplines will be appropriate here:

Civil engineering is acknowledged as the oldest of the contemporary engineering fields. Initially the term reflected a distinction between military and civilian construction. In a word, civil engineers are concerned with infrastructure – buildings, roads, bridges, canals, water supplies, and railroads, among many other things. Structural engineering and environmental engineering are often viewed as sub-specialties within civil engineering; many academic departments now carry the name "Civil and Environmental Engineering." All of us interact with the products of civil engineering every day, when we use public roads and sidewalks, when we buy things shipped over long distances, and when we draw water from a faucet.

Mechanical engineering is concerned with the management of force and the movement of energy, especially as they involve machinery. This simple definition belies an extremely diverse field, which includes sensors and actuators, manufacturing processes, rotating equipment (e.g. pumps, compressors and engines), propulsion, heating and cooling, robotics and steam power generation. Mechanical engineers work with the flow of fluids, so that aerospace engineering (which has a strong interest in the way that air flows around airplane parts) is considered a sub-specialty; academic departments of "Aerospace and Mechanical Engineering" (or in the opposite order) are common. We encounter the products of mechanical engineering in our home appliances, home heating, cooling and water systems, and vehicles.

Chemical engineering covers the use of chemical reactions and complementary processes (e.g. distillation, filtering, drying) to produce a useful result, typically at commercial scale. Very few products come to us in their original forms; they must be synthesized, purified, combined, etc. from a relatively small set of raw materials. Waste products from production processes must often be treated, and discarded products are often recycled. Chemical engineers work on the production of fuels, food, drugs, agricultural chemicals and consumer products, among others. Chemical engineering is making increased use of biological processes so that now some academic departments are being called "Chemical and Biological Engineering" to emphasize biological routes for synthesizing chemicals or "Chemical and Biomedical Engineering" to emphasize such areas as design of medical devices.

Electrical engineering combines electric power (including electric machinery), communications and computing. All, of course, are ubiquitous in our daily lives. Sub-specialties include display devices and electro-optics, biomedical applications, and radar. In the developed world we don't think much about electric power until we don't have it, but behind the wall switch is a lot of technology for its production, transmission and use. "Power Electronics" covers specialized large semiconductor devices and circuits which facilitate the transmission and use of power. Communications, especially digital communications, and computing get more attention, with many academic departments now called "Electrical and Computer Engineering". Computer engineering emphasizes hardware, whereas computer science and software engineering emphasize software, but this is a fuzzy boundary.

Industrial engineering concerns itself generally with large-scale processes such as used for manufacturing and distribution (logistics). An industrial engineer might, for example, design a manufacturing plant around requirements for production rate and flexibility, and then specify or design the automation system that directs the movement of materials around the plant. The industrial engineer may play also an important role in plant operations, maintaining production quality, managing raw material supply, setting personnel schedules and ensuring safety and environmental compliance. Some industrial engineers work outside of industry, such as in health-care settings where they help to design facilities for optimal patient care at minimum cost. There is a considerable overlap between industrial engineering and operations research, with the latter typically considered a business discipline along with finance and marketing. There is also an overlap with economics.

For the most part, other engineering disciplines borrow from, specialize, and expand upon these five. For example, pulp and paper engineering is primarily a chemical engineering discipline but also includes elements of mechanical engineering. Similarly, agricultural engineering benefits from mechanical engineering (design of farm equipment), civil engineering (agricultural infrastructure and waste treatment) and chemical engineering, especially its biological component.

A new engineering school

In my opinion the traditional arrangement of courses tends to promote narrow thinking, and convinces a student studying in one discipline to think that he can't easily move into another. This isn't merely a perception; dueling course requirements and prerequisites often do make it difficult to switch, especially among specialized courses of study such as accredited engineering programs. This siloed thinking also, in my view, doesn't necessarily prepare a student for what's coming. A narrow focus can create unnecessary difficulty for graduates in moving between fields. Here's an alternative: Once the basic science and math courses have been covered, build the undergraduate engineering curriculum around themes common to a number of fields. These themes would for some students replace traditional majors, and might include for example:

Materials Science (failure analysis, corrosion and electrochemistry, polymers, semiconductors, plasma phenomena, cryogenics, x-ray diffraction and other technologies for visualization)

Machinery (pumps, motors and generators; food-processing machinery, laboratory automation, motion control and robotics, risk and failure analysis)

Manufacturing (supply chain management, occupational health and safety, machining, robotics, semiconductor processing, food and pharmaceutical processing, casting, quality control)

Instrumentation, Controls and System Dynamics (measurement of physical properties, feedback and stability, chemical process control, structural resonance and seismic response, mechanical and other actuators, motion control and robotics, system science)

Statistics and Signal Analysis (data science and analytics, statistics, noise reduction, signal filters, image processing, design of experiments)

Energy (thermodynamics; chemical, electrical, mechanical, pneumatic, hydraulic transmission of energy; energy conversion and heat transfer; energy storage, social and environmental issues)

Flow phenomena (gas, liquid and multiphase flows; aerodynamics, discrete flows e.g. vehicular traffic and queuing systems; control systems)

Engineering Economics and Operations Research (data science and analytics; macroeconomics and finance; software design and implementation; quality control and reliability; investment analysis, project management, supply chain logistics and optimization)

Even specialty-specific courses can be interdisciplinary in their own ways. An electrical engineering course in transmission lines, for instance, could cover fiber optics, radio-frequency transmission and power-frequency systems (transmission and distribution), as well as some acoustics. The mathematical fundamentals for these different subjects are not themselves so different, but students, particularly under-graduates, may never learn this.

Such programs could better prepare their graduates for careers in engineering than the narrowly-specialized traditional programs. Programs in engineering science do exist today, however they are oriented more toward scientific fundamentals than engineering practice. For the most part they put together existing engineering, science and math courses in different combinations than do traditional curricula. There's nothing wrong with that, or with conventionally arranged engineering curricula, but what I'm describing here is a little different. And, unfortunately, an undergraduate program organized along the lines I'm proposing here might not easily receive accreditation or be recognized by state licensing authorities as equivalent to a traditional undergraduate program. It might therefore be necessary for a graduate of one of these programs to go on to a traditional master's degree, or acquire some years of experience beyond the usual minimum, to meet educational requirements for licensure. The potential good news, however, is that these graduates may be better prepared for lifelong careers than some of their conventionally-educated peers.

Are four years enough?

A graduating high-school student needs four years, nominally, to acquire the necessary training to start a career in engineering and meet the academic requirements for Professional Engineer licensure. The student's freshman and sophomore years are devoted to acquiring the background in science and math needed to succeed in the engineering courses taken in the junior and senior years. Having completed an accredited program, the young engineer will also have to accumulate some years of experience and pass a test to become licensed. Experience beyond the minimum will usually be necessary if the engineer's academic program was related to engineering but not accredited.

We don't expect our physicians, lawyers, dentists and veterinarians, among others, to do anything like that. They are first educated and once educated are trained, in a process that in the aggregate takes more, often considerably more, than four years. Yet engineers also have a lot to learn, and engineering is changing at least as quickly as any of these other fields. (Whereas fields of engineering are always changing with advances in technology and science, physicians don't often have to deal with newly-discovered organs*, and the practice of law tends to change by evolution rather than discontinuity.) I've seen this personally – lab equipment that I used as an undergraduate was by about 30 years later in a display case being exhibited as historical artifacts. To keep up with changes in the field, instruction on established tools and techniques must give way to the subjects of current research interest and to career opportunities seen as most promising. While it's of course desirable for a newly-graduated engineer to be up to date, there just isn't enough time for everything, so

---

* In late 2020 researchers discovered, by accident, heretofore unknown salivary glands

topics considered less than current must be de-emphasized or dropped entirely. In some cases, subject areas that are economically important but don't attract much research money are dropped to make room for areas considered more timely, and perhaps more lucrative for the institution.

The American Society of Civil Engineers – representing a branch with a comparatively high proportion of Professional Engineers – has gone on record as saying that a master's degree should be the minimum educational requirement.[60] In 2006 the National Council of Examiners for Engineering and Surveying (NCEES) recommended that by 2020 a masters degree or equivalent be required of new licensees. The recommendation was dropped in 2014[61] but the NCEES would still like to see a requirement for education beyond the baccalaureate[62]. Over time, and with the concurrence of other professional societies, a mandate for additional formal education might help improve the professional standing of engineers. The various engineering disciplines must stay in synchronization with one another, however, lest one branch of engineering become more "professional" than another. But it won't happen overnight, and I don't think it will happen at all until we resolve the linguistic ambiguities discussed in the previous chapter. Unless done very carefully, a requirement that a prospective Professional Engineer have a masters degree may for many simply impose the direct and opportunity costs of spending another year or two in school, without a corresponding increase in earnings. One hopes that any additional educational requirements will help to improve engineering practice and elevate engineering to a true profession, not merely help fill classes at educational institutions. With educational costs rising more quickly than the general rate of inflation it will take a larger part of the engineer's career to break even. Further, employers in exempt industries might be reluctant to hire someone with a master's degree to fill a job that the employer believes does not require the additional education or demand additional compensation.

Conceivably also, if such a move were to materially increase salaries for some Professional Engineers, it could perversely reduce job opportunities for others as employers move as much work as possible to unlicensed personnel, including those residing overseas. There is a close parallel in the field of physical therapy. In the past, physical therapists (PT's) were generally required to have master's degrees. Today, the requirement for entry into the field is the Doctor of Physical Therapy degree, DPT. Programs at the master's degree level aren't even offered in the United States any more.[63] This hasn't necessarily been positive for the career interests of PT's. Some complain that any additional pay they receive isn't justified by the costs associated with the additional education. Employers have an incentive to replace as many licensed PT hours as they can with the time of lesser-trained individuals, i.e., Physical Therapy Assistants with a minimum of two years of education after high school[64] or physical therapy aides, generally needing no training beyond high school and no license.[65] A similar situation exists in the field of pharmacy, where employers can now hire pharmacy technicians to supplement licensed pharmacists, who now must have doctorates.

Continuing Education

Continuing education is one way for the engineer to acquire new skills and keep up with technical change. Opportunities for informal continuing education are plentiful – they include routine on-the-job learning, websites, blogs and hardcopy publications covering your field, online courses (some free), professional organization meetings and trade shows. More formally, continuing-education courses are organized and presented by employers, professional societies, universities and independent training companies.

State licensing bodies may require formal continuing education as a condition for license renewal. This brings up the question of accreditation. The International Association for Continuing Education and Training (IACET)[66] accredits training providers, whose courses then qualify for IACET Continuing Education Units (CEU's). But the IACET doesn't own the term CEU, so others are free to offer their own

CEU's without regulation by any third party. Your state licensing board may also grant credit for university courses not otherwise accredited or presented as part of a recognized academic program. If you have a PE license and are subject to continuing-education requirements, make sure that any courses you take will be recognized by your board.

Five-Year Programs

For those who wish to get trained to a professional level and educated through the same program, five-year courses of study offer some options. One may spend five years as an undergraduate and get two degrees, usually a BS in engineering and a BA. This sometimes involves attending two different colleges: The student first spends three years at a liberal-arts institution and then two at an engineering school. At the first institution, the student completes the first two years of a typical engineering school curriculum focused on math and science, as well as most of the selected liberal arts curriculum. Then the student moves to the engineering school to spend the next two years focusing on specialized upperclass engineering courses. These combined programs are prearranged, either within one institution such as Rutgers University or Lehigh University, or between institutions. As an example of the latter, Columbia University has "Combined Plan Program" agreements with almost 100 other institutions. Admission to the two years of engineering specialization may not be guaranteed; make sure you understand the requirements of any program you're interested in. Such a program, at the cost of an additional year of college expense and the opportunity cost of deferred income, can provide a good basis for moving into a field which isn't strictly technical but for which a solid technical background can be helpful. Corporate management is an example.

This option would be less attractive for someone who plans to spend his career doing engineering, as with a similar investment in time and money he could receive a master's degree in an engineering branch or a master's in engineering management (MEM).

A related option involves staying on the same campus and getting two degrees in (usually) related subjects, such as in electrical engineering and engineering physics (Lehigh University), or computing and cognitive sciences (University of Pennsylvania). Some combined programs can be completed in four years, but check accreditation status, course load, and whether year-round attendance is required.

If financial aid will be a necessity for you to attend college for four years, it can be even more of a necessity for five. Some colleges may make special provisions to ease the cost of adding a fifth year, but this is something you should be sure of before committing. Find out what would happen if you got partway through a combined program and found you couldn't afford a fifth year – would you be able to get an engineering degree after only four? Would it be accredited?

Some day there may again be a military draft. The United States hasn't conscripted people for military service in a couple of generations but men, including non-citizens, are still required to register with the Selective Service System. If a future draft operates under rules similar to those in effect during the Vietnam War, college students will be able to defer military service for up to four years, but would then be subject to being drafted before being able to complete a fifth year.

A five-year program could be a good choice for a student who develops an interest in engineering after beginning undergraduate studies at a school that does not offer engineering. As with any engineering program, prerequisites may come with their own prerequisites. The earlier that one decides to investigate such a dual-path undergraduate education the less likely he is to end up with unusable credits or unsatisfied prerequisites. Science and math courses will be available in a liberal-arts program but may be more oriented toward majors in their fields, rather than to practical applications.

Should you get a Master's Degree?

Surveys generally show that engineers holding master's degrees of one kind or another earn greater annual salaries than their peers with only bachelor's degrees. A variety of programs are available. However the holder of a master's in engineering or a related field (collectively, a "technical master's") is likely to earn less than the holder of an MBA.

The best known of the technical master's degrees are the Master of Engineering (M.Eng. or similar) and Master of Science (M.S.). An institution may offer one or both. For either, normally the student brings a bachelor's degree in a field of engineering, and finishes with a master's in the same discipline. But there are differences in requirements and outlook between the two. The M.Eng emphasizes engineering practice, and may include more coursework than the M.S., including some business courses. There may be a final project or report required for the M.Eng, but not a formal thesis. It is considered a terminal degree; the M.Eng graduate usually does not plan to go on to additional formal study in the same engineering field. In contrast the M.S. degree emphasizes research, and while it may require less coursework than the M.Eng., it usually requires a thesis representing original research. It is typically expected that, having received the M.S., the student will move on to a doctoral program. Programs leading to the M.Eng. may be administered by an institution's engineering school while M.S. programs are administered by its graduate school.

Of course there are variations on these themes. It is possible, for example, to get an M.S. through a Professional Masters program, following a course of study that would at other schools lead to an M.Eng.

Costs aren't necessarily the same for the two types of program, and generally there is more financial support available to M.S. students in the form of research and teaching assistantships. These assistantships are usually for professors-in-training and are not so often available to those pursuing the M. Eng., although an M. Eng. student may have to pay for fewer semesters on the way to a degree. Funding from outside organizations may be more readily available to support student research than to prepare students for careers in engineering practice.

Off campus, there doesn't seem to be a lot of difference between the way that M.Eng. and M.S. degrees are viewed. An engineer entering the job market after graduating with an M.S. degree should not encounter skepticism from employers.

For an individual with a bachelor's degree not in engineering, a masters degree can be the first engineering degree. The Late Entry Accelerated Program (LEAP) program[67] at Boston University, as one example, is designed for students with non-engineering undergraduate educations. Schools without such programs may still be able to accommodate non-engineering graduates on a case-by-case basis. Additional prerequisite coursework will usually be necessary, the amount depending largely on the student's undergraduate major. Those with undergraduate majors in physics, chemistry or mathematics will likely need less prerequisite coursework than, say, English majors.

There are also programs leading to the Master of Engineering Management (MEM) degree, or a degree with a similar name, like Master of Science in Engineering Management. This degree is intended for those who want to pursue technology-oriented careers with an emphasis on management. Different from the M.Eng. and M.S. degrees, the MEM is not specific to an engineering branch, like Civil Engineering; rather, engineers from different disciplines go through the same program and get the same degree. This degree is often compared to the MBA, but whereas an MBA program may accept students from any undergraduate discipline, the MEM is primarily for engineering graduates. MEM students are typically directing their careers toward group technical leadership, and departmental and project management, but not necessarily toward general management, for which an MBA would be more appropriate.

MEM programs typically include subjects from operations management (overlapping with Industrial Engineering / Operations Research and the quantitative methods courses in MBA programs), information technology (also often with some commonality with courses in MBA programs) and general management. They may also include a requirement for some traditional engineering coursework. It is often possible to complete an MEM degree within one year, especially in conjunction with a formal BS/MEM program as offered at some schools. Courses needed for the MEM degree can largely be taken online, since laboratory facilities are usually not needed as they are with many traditional engineering subjects. A large number of online-only MEM programs are available. The MEM thus becomes an attractive choice for engineers who wish to further their educations at relatively low cost and/or while working full time.

The Master of Business Administration, the MBA, is normally a two-year program offering concentrations in such areas as finance, corporate strategy and operations. Engineers often pursue MBA's as preparation for supervisory or project management, or to completely leave engineering with a move to general management. Some business schools including Cornell's Johnson School of Management, Northwestern's Kellogg School of Management, and Notre Dame's Mendoza College of Business have begun to offer one-year programs instead of or in addition to their traditional two-year programs. These programs are intense, requiring a full calendar year of study to earn the same degree that normally takes two years. Their advantages, of course, are financial – reduced direct costs as well as opportunity cost. MBA programs, particularly prestigious ones, are notoriously expensive.

The "STEM MBA" is an MBA degree with an emphasis on one or more of the applicable fields – science, technology, engineering or mathematics.[68] These programs were developed not to serve Americans, but primarily to attract foreign students. By receiving the STEM designation from the Department of Homeland Security, which has been possible since 2016, an MBA program can offer international students the likelihood of an additional two years in the United States through an extension of post-graduation Optional Practical Training.[69] Notwithstanding the origins of these degrees, they offer a valuable opportunity to American students.

Those interested in moving toward research, without getting a doctorate degree, may also be interested in one of the emerging Professional Science Master's (PSM) programs.[70] The largest single area of PSM programs is biology, with environmental sciences second.

Chapter 3: Starting Your Career

Evaluating Job Offers

If you're fortunate enough to receive more than one attractive job offer, review the financial aspects of each carefully. Beyond salaries, are there bonuses? Companies are increasingly turning to bonuses as tools that will allow them to increase, or if they choose decrease, the money they pay you. If you get a raise of some percentage then your next raise will be calculated from that, but a bonus grant one year won't necessarily influence your salary or bonus the next. In the opposite direction, a decrease in bonus amount from one year to the next may be better received than a decrease in salary.

If the offer includes stock options, find out the vesting schedule; you will probably have to work for the employer for a number of years before all of the options offered actually become yours. Try to find out the recent cost history of the employee portion of health care costs (basic coverage costs, co-pays, and deductibles) as they can go up faster than your salary. Similarly, learn the history of the employer match to your 401(k) contributions – what it is now and whether it has gone up or down within the past few years. The best way to get this information may be to ask open-ended questions of your prospective peer colleagues.

If you have any specific medical or dental needs, like the need for an expensive drug for a chronic condition, find out how well they're met by the employer plan.

If your company offers you life insurance, compare its present and likely future costs to what you can obtain on the general market.

Some engineers find in startups an opportunity to be exposed to aspects of technology and business well beyond what they were hired for. You can learn a lot working for a company that has more job descriptions than people. But then keep your eyes open as the company brings its products and services to market, grows, and becomes more bureaucratic; frequent and informal sharing of information may come to be replaced by formal management meetings to which you may not be invited. And your continued growth within the company may depend more on how you're perceived as potential "management material" than on your contributions as an engineer. If you go to work for a startup, then, be sure to measure your career progress against the progress of others.

Every company has a unique and individual culture. Exciting new companies get the attention, but old-line companies can also aggressively innovate. Even stable companies in established markets have to adapt to new technologies, competition and opportunities. And starting at one type of company won't preclude your moving to another; wise managements appreciate the cross-fertilization that can come from bringing in people with a variety of backgrounds

Supply and Demand: Is there a shortage of engineers?

This discussion is here, in the section of the book that covers the beginning of an engineer's career, because I suspect that many high school students are motivated to become engineers by a barrage of news articles and pronouncements by public officials about how dependent America is on technology, how we're in danger of falling behind our international competitors and adversaries, and how there are never enough engineers to fill available jobs.

Predictions of engineering shortages so great as to jeopardize our economy and security, or at least implying high demand for engineers, have continued from time to time to this day. The presumed existence of a long-term engineering shortage drives the acceptance of thousands of H-1B applications every year. But is there really such a shocking shortage?

There are indeed times when engineers are in short supply. As World War II led to the Korean War, and the Korean War led to many years of Cold War between the United States and the Soviet Union, engineering enrollments were low and demand high. As Time Magazine reported on April 21, 1952, "Another [employer] promised the University of Santa Clara to employ even those engineering students who flunk their finals."[71]

And then, on October 4, 1957 in the midst of the Cold War, the Soviet Union launched Sputnik, the first artificial satellite, stunning the American people. Less than a month later the Soviets launched a second, larger Sputnik, this one carrying a dog. The first two American attempts, using the Navy's Vanguard satellite, failed at or shortly after launch, but then on January 31, 1958 the Army's Explorer was launched successfully and truly put the United States into the space race.

The "Sputnik scare" shook the US out of its belief in the near-automatic superiority of its science and engineering, and validated the Soviets' ability to threaten the US with missiles carrying nuclear weapons. As history shows, the US did successfully respond to the challenge represented by Sputnik, eventually landing men on the moon and bringing them back safely, a feat that no other nation has since accomplished. To achieve this the US had to motivate large numbers of young Americans to pursue careers in engineering.

Did then the talent shortages exposed by the Sputnik scare improve the economic position of American engineers and make the profession more attractive? The answer is mixed. The McNamara–O'Hara Service Contract Act of 1965[72] (SCA) prohibited "wage busting" by government contractors. Wage busting is the practice of reducing employee wages and benefits from current values to make a proposal for a new contract more attractive and competitive. The SCA explicitly excluded coverage for personnel working in a "bona fide ... professional capacity" which included engineers. The apparent presumption was that professionals could negotiate for themselves and so were not in need of statutory protection. With engineers ineligible for legal protection, their salaries became an obvious target for contractors preparing competitive bids. The situation became so difficult for the government that it was forced to issue "special procurement procedures" to provide some protection to engineers, as described in a report submitted in 1978 by the Comptroller General of the United States.[73] A reasonable person might conclude that the contractors' action in pushing down engineering wages furnished good evidence that, in that area and that time, there was not a shortage of engineers.

Spot shortages of engineers, in a given region and/or specialty, can happen at any time and for any number of reasons. So can surpluses. As an example, fully electric automobiles now account for a small fraction of auto sales in the United States. But it has been reported that development of gasoline engines for automobile use has effectively ceased, in favor of electric drive trains. This shift has probably led to a surplus of engineers with expertise in gasoline engines and a shortage of those experienced in electric drive trains. To resolve the latter, manufacturers can hire new graduates and have them specialize in electric drive trains early in their careers. Those who worked on gasoline engines, especially older engineers, could find themselves in line for layoff, while younger ones try to get retrained or start working on their applications to business school.

A free labor market should be able to adjust itself to changes in supply and demand; this is in fact the fundamental purpose of such a market. Granted, the market's response will not be immediate; there will be a considerable delay as it takes at least four years for a high school graduate to become a qualified entry-level engineer. However, this is hardly the only case where there can be a long delay time between an increase in demand and the corresponding increase in supply. The assertion that a free-market society can experience a sustained decades-long shortage of employees trained for a particular occupation, especially an occupation generally regarded as attractive, is baffling.

Of course, the mechanism through which free markets match supply and demand is price, and it is certainly possible in a free-market economy for employers to decide that the quantity of capable employees willing to work for the wages they wish to offer is inadequate. Thus we can view the claims about a persistent US engineering shortage to be more accurately of "a shortage of engineers willing to work at wages that employers would like to pay" or even more to the point "a shortage of engineers willing to work at wages that would be readily accepted by foreign students presently studying in the US and who would like to stay here after graduation".

There is ample evidence that there has been no large-scale shortage of engineers for many years:

The unemployment rate is not unusually low. If engineers are in short supply we'd expect their unemployment rate to be lower than for workers with four-year degrees in other fields. It's difficult to draw conclusions from official unemployment figures, since these numbers don't include part-time workers, the discouraged workers who have stopped looking (technically, left the labor force), and those who would otherwise count as employed engineers but were forced to change fields and found employment, often underemployment, elsewhere. For the most part, though, official statistics show unemployment among engineers as being in line with that of other groups having similar education levels.

Engineering salaries have not gone up at an unusual rate. A sustained low supply/high demand situation should result in raises routinely higher than average, particularly after the first few years of employment. I have seen no reports that this is happening, although there may be exceptions in "hot" fields. Such exceptions are likely temporary. Particularly in computer-oriented fields, technologies rapidly move into and out of fashion. (Fortran programming, anyone?) Rapidly-changing job requirements tend to depress salaries for all but early-career engineers.

Non-practicing engineers are not, in large numbers, abandoning their careers in sales, marketing or management to return to hands-on engineering. We would expect this if engineering were more attractive than these other fields. Surely it does happen on occasion. I'm not aware of any studies that quantify this phenomenon but my observation is that the number has to be quite small. Once an engineer moves out of engineering it's unusual for him to come back. It has been estimated that about half of all graduate engineers are working in other fields. This suggests a shortage of suitable jobs, not a shortage of suitable people.

Mid-career professionals in other fields are not abandoning their jobs to study engineering. There are always some adults studying to become engineers, but in my observation these are mostly people trained as technicians, often in the military, for whom becoming an engineer does not constitute a career change so much as a path for advancement in their present occupational fields. It appears to be more common for people trained in other fields to learn programming and become application developers. "Learn to code" has become an epithet, suggesting that someone's current occupation is in decline.

Early- and mid-career engineers often seek other options. Dropout rates from engineering schools are typically high, but individual decisions to leave the field don't end at college graduation. Graduate engineers often pursue MBA's and other nontechnical degrees, whether soon after graduating from engineering school or after working for a few years. It's axiomatic in marketing that the easiest customer to acquire is the one that a company already has. Similarly, the people easiest to recruit into the engineering profession should be those who already have engineering degrees, yet many choose to leave. If their employers are experiencing shortages of engineering talent, these employers don't always seem to be investing a great deal of energy into retaining, and if necessary retraining, the talent they already have.

Consolidating the two points above, one notes that it's easy to find technical sales personnel, business managers, financial planners, lawyers, physicians and so on who used to be engineers, but virtually impossible to find engineers who used to be technical sales personnel, successful business managers, financial planners, lawyers or physicians. (A near exception that helps to prove the rule is the case of Irwin Jacobs, who entered the School of Hotel Administration at Cornell University and then switched to electrical engineering. He eventually helped found the chip manufacturer Qualcomm.)

Support staff is generally low and decreasing. Engineers typically answer their own telephones, organize their own expense reports, edit their own reports, and so on. If there were truly a shortage of engineers, corporations would hire support staff to do those portions of the engineer's work that don't require his capabilities, reserving their limited engineering resources for tasks that only an engineer can perform. As we know, that's not what happens. In contrast, the medical industry, as one obvious example, is traditionally organized around minimizing the time that a doctor spends with each patient, in favor of receptionists to greet patients and make appointments, staff to take medical histories, phlebotomists to draw blood, various specialized technicians to run X-ray machines and such, and direct support personnel including nurses, nurse practitioners, and physician assistants. (A doctor might tell you, though, that she spends far too time on required paperwork.) Similarly, a modern pharmacy may have only one pharmacist on duty at a time, with most of the work being done at lower cost by pharmacy technicians.

Age discrimination would be non-existent. Employers would be happy to hire or retain any qualified engineer, regardless of age.

Employers would be more willing to invest in continuing education. There would be a premium on retaining and retraining existing talent, rather than waiting for a new college graduate to come in with detailed knowledge of the latest-and-greatest techniques, software applications, and so on.

Employers eschew other investments that could also improve engineering productivity. If engineers were in short supply and employers truly wanted to help improve engineering productivity, they wouldn't put engineers into "cube farms" arranged so that a conversation in one cubicle can be heard four cubicles away. And some cube farms have been replaced by long tables. This may be called "encouraging collaboration" although face-to-face collaboration is often forced into break rooms and corridors. Even collaboration by telephone becomes difficult as one conversation can disturb several nearby workers. The forced physical distancing and working from home necessitated by the COVID-19 pandemic may someday be found to have led to an increase in job satisfaction and productivity.

Considering all of the market signals that would if present evidence a shortage of engineers, it is clear that no overall shortage exists. However the absence of a broad shortage of engineering talent doesn't preclude the occasional emergence of spot shortages, as has been noted. That said, we can expect our educational institutions, their students, and engineers early in their careers to move quickly to fill any such gaps.

Researcher Michael S. Teitelbaum counted five boom-and-bust cycles in the employment of engineers and scientists between the end of World War II and 2014.[74] The Sputnik scare initiated cycle number two. Each cycle began with an alarm – a belief that an inadequate supply of such workers could compromise our security or constrain our economic growth – followed by the encouragement of a large number of students to enter the field, and then by a bust as demand and funding stabilized or declined. Each cycle spanned about 10 years. We engineers recognize this as oscillatory behavior, which can be caused by too much feedback; that is, trying too hard to correct a problem. This tendency to oscillate is reinforced by the years required between the beginning and completion of the necessary education. The analogy isn't perfect, since each cycle has primarily affected a different set of specialties. Still, we could do a lot better, probably not eliminating these oscillations but at least dampening them.

It also appears that there are occasional shortages of entry-level engineers with BS degrees. Employers sometimes claim that they have an inordinate amount of trouble filling such positions. One reason seems to be that foreign engineers, already holding BS-level degrees, often come to the US on student visas and go directly to graduate school. Thus they don't participate in the entry-level segment of the engineering labor market. U.S. engineers may also decide to go to graduate school, or change fields and go to business school, law school, etc. They may also go directly to business-oriented positions because these pay better. So the number of newly-graduating engineers available for entry-level engineering jobs may be smaller than we would expect. The suggestion here is that engineering salaries, even for entry-level personnel, have not risen to the market-clearing price.

We must also consider the effects of age discrimination: The existence of a strong demand for younger engineers doesn't preclude a weaker demand for mid-career and older engineers. The labor market doesn't offer enough mid-career jobs to absorb all of the early-career engineers wishing to move up, or at least it doesn't have enough attractive mid-career jobs to encourage large numbers of people to stay in the field. If our society does have a persistent shortage of engineers, it has failed to utilize what is probably its most powerful tool to address it–retention of practitioners who are already trained and qualified.

Alternative career paths

Is engineering the new liberal arts? The chairman of a major technology company thinks so.[75] Once ensconced within a corporation, the young engineer has to decide whether to try to build a career in engineering or seek growth and fulfillment doing something else.

Some of today's most prominent Americans were trained as engineers before moving on. Examples are entrepreneur and three-term New York City mayor Michael Bloomberg (electrical engineering, Johns Hopkins University, Amazon founder Jeff Bezos (electrical engineering and computer science, Princeton University, musician and songwriter Tom Scholz (mechanical engineering, Massachusetts Institute of Technology, musician Herbie Hancock (electrical engineering, Grinnell College, and baseball player and coach Joe Girardi (industrial engineering, Northwestern University. Alex Padilla, selected to take the Senate seat vacated by Vice President Kamala Harris, is a graduate of MIT in mechanical engineering. ESPN sports analyst André Snellings holds a BSEE from the Georgia Institute of Technology and a doctorate in biomedical engineering from the University of Michigan. American leaders now deceased include the movie director Frank Capra (electrical engineering, California Institute of Technology, artist Alexander Calder (mechanical engineering, Stevens Institute of Technology, and U.S. President Herbert Hoover (civil engineering, Stanford University. And this phenomenon isn't limited to Americans. Brits include actor Rowan Atkinson ("Mr. Bean", electrical engineering, Oxford University and the late movie director Alfred Hitchcock (marine engineering, London City Council School of Marine Engineering).

Engineers are trained to solve problems, a helpful skill in just about any field. Further, in engineering we accept that some answers are right, some are partly right (might work, but with downsides) and some simply wrong. We may therefore be inclined to be more objective than some when evaluating a situation and its possible solutions. While there are adults who do begin the study of engineering after working in other fields, the dominant direction of exchange between engineering and other career fields is from the former to the latter. Many who trained to be engineers have found venues more attractive than engineering in which to use their engineering skills. And a knowledge of engineering principles and methods should be helpful in any organization that develops or uses technology, which is of course all of them.

Engineering as a field of study is thus arguably more attractive than engineering as a career. There may not be an obvious connection between, for example, the study of engineering and success as an actor, but it is easy to see the connection between engineering and Calder's kinetic mobiles. Engineers who leave the field, especially those who become prominent in their new fields, sometimes make lighthearted comments about not being able to do the work, either in school or later. Although we don't often hear "I got laid off from several engineering jobs so then decided I'd better go to law school" or "I got a doctorate in engineering and couldn't make any money, so I decided to move into finance," in private conversations I've heard both statements and more. Everyone who leaves engineering has his own reasons, but I don't think that an inability to do the work is high on the list. Not having work to do, i.e. being laid off, is more common. Possibly the most common reason is the simple conclusion that one's engineering career has played itself out and that further progress will require a new career, often but not always using engineering as a foundation for the next pursuit.

As already noted, engineers often move into management. Much of the time their first nontechnical jobs are either as project managers or first-level managers supervising other engineers. That's where many stay, while they see their peers without technical backgrounds, from operations, sales, finance and so on, move further up the ladder. So engineers who wish to move further into nontechnical positions will have some work to do. In my experience, for those who want to leave engineering, sales is a good path out. If you're selling a technical product or service, your deep understanding of what you're selling should be a big advantage. Appropriate credentials, i.e., an MBA, will of course be helpful as well.

Chapter 4: Career Issues

A global profession

Many would agree that the major career issue facing America's engineers today is outsourcing, or more accurately, offshoring; that is, the movement of jobs once held by American engineers, to competitors in lower-wage countries. This may also include the creation of new jobs overseas in preference to doing so here. American companies in the past often sent manufacturing offshore while keeping engineering in the US, but offshoring of American engineering jobs has now followed.

To the corporation, the logic in favor of offshoring engineering work is compelling: The mathematical, physical, chemical and biological processes that ultimately govern the work of engineers are the same everywhere. If the work that engineers do can be seen as fungible; that is, as a commodity, then there is no reason not to simply send it to be done by the lowest bidder, who will most commonly reside in a low-wage country.

The debate over sending engineering jobs offshore, to the extent that there is still such a debate, would benefit from corporate America's honest declaration of its motivations. But the debate is often obscured by comments about the number of American engineering graduates compared to greater numbers of graduates from other countries. A common theme in management-oriented writing about American engineering can be summarized as, "Not enough American students want to study engineering. Therefore, we have to send more engineering jobs overseas, and as well open up more visas for foreign engineers to come here." But if the labor market in the US were allowed to work the way markets are supposed to work, domestic supply of engineers, to the extent that it is a constraint, could over time catch up with demand. The first "new" source of talent would be discouraged mid-career engineers who are now leaving the field.

Immigration and the American engineer

By the time the United States went from British colonial rule to independent nationhood, engineering was well established elsewhere in the world. Sophisticated machines had been built in ancient times, mainly by the Greeks and Romans. Inventions included weapons, water pumps and steam engines. Romans built aqueducts more than 2300 years ago. There are claims that in the 15th century, China had wooden ships that were over 400 feet long, and that in 1745 the first oil refinery was built in Russia. In the early 1800's, as the United States was establishing itself as a nation, Europeans were building electric motors. In 1812, Francis Cabot Lowell returned from Great Britain having memorized the workings of economically sensitive textile machinery, allowing him to establish a competitive industry in the United States.

Initially, then, it is clear that the United States was an importer of technology. Over time, this changed as Americans pioneered the invention and development of telegraphy and telephony, lighting, airplanes, automobiles, integrated circuits, and so on. In many cases these Americans were immigrants, such as Alexander Graham Bell, inventor of the telephone, who was born in Scotland; or had studied abroad, such as Robert Fulton, who worked in France and England before returning to the United States to build the first successful steamboat service.

The United States has continued to benefit greatly from foreign-born and foreign-trained engineers and inventors. In more recent times, these have included Ralph Baer, inventor of the video game (Germany), Sergey Brin, co-founder of Google (Russia), Luther George Simjian, inventor of the automatic teller machine (Armenia), Jerry Yang, co-founder of Yahoo! (Republic of China, a.k.a. Taiwan), and C. K. N. Patel, developer of the carbon dioxide laser (India).

A complete list, if the compilation thereof were even possible, would clearly fill a book. Doubtless, for the health of the American economy and thus the defense and prosperity of its people, inadequate immigration would be detrimental. But can there be such a thing as excessive immigration? On an axis that goes from too little immigration to too much, where do we place ourselves now? And wherever that is, are we making the best choices with respect to the people who come in?

Immigration starts with education, often at the secondary level or even below. Students from wealthy families overseas often pay full price for their post-secondary education here in part because a US education can lead to a visa to stay. From the perspective of US students this has costs and benefits: The presence of a large number of international students helps improve the financial stability of a college but at the same time reduces the number of seats available for Americans. The ability of these wealthy families to pay top dollar may increase the cost for American students who don't qualify for financial aid, because college "sticker prices" tend to track the money available. US colleges must, of course remain open to foreign students, particularly from friendly countries, but it's possible that their numbers have been too high.

The H-1B visa is the principal means by which foreign engineers come to work in the country. The visa's history goes back to the Immigration and Nationality Act of 1952,[76] passed over the veto of President Harry Truman. That act made specific provisions for immigrants "whose services are determined by the Attorney General to be needed urgently in the United States because of the high education, technical training, specialized experience or exceptional ability of such immigrant..."

Today's H-1B is for temporary employment of individuals in "specialty occupations"[77] as well as certain fashion models of "distinguished merit and ability". The H-1B must be applied for not by the engineer but by the employer "sponsor"; thus, job ads with language like "sponsorship not available" indicate that the employer is not willing to invest the effort and expense required for sponsorship, or that perhaps for security reasons the job is open only to US citizens and permanent residents, the latter informally known as "green card" holders. The H-1B visa is nominally available to a wide range of professionals and others in "specialty occupations", but engineers, especially software engineers, dominate the category. Under current law most employers (those not deemed "H1-B dependent" or "willful violators") are not obligated to first seek to hire US citizens and residents; these employers can go directly to the H-1B process.

The number of H-1B visas available in a year is mostly set by law. On October 17, 2000, President Bill Clinton signed the American Competitiveness in the 21st Century Act of 2000[78] raising the visa cap to 195,000 for fiscal years 2001, 2002, and 2003. It exempted from the cap individuals employed by universities and affiliated research institutions and made several other significant changes. On signing the bill the President said in part, "Today, many companies are reporting that their number one constraint on growth is the inability to hire workers with the necessary skills." On that day, the NASDAQ Composite closed at 3,213.96, down more than 36 percent from its peak of 5,048.62 reached in March of that year. The tech bubble had long since burst, but evidently not everyone in the country noticed.

In the fiscal year 2000 (October 1, 1999 to September 30, 2000), the Immigration and Naturalization Service (INS, since replaced by the Bureau of Citizenship and Immigration Services, Department of Homeland Security) approved 136,787 H-1B visas.[79] Of these, 74,551 (over 54 percent) were "computer related" and another 17,086 (12.5 percent) were in "architecture, engineering and surveying", for a total of roughly two thirds.[80] These counts include both new and extended visas.

The number of bachelor's degrees in computer science granted in the US in 2000 was 37, 519;[81] the number of master's was 777. In engineering, the number of bachelor's degrees was 59,487[82] and there were 5,384 master's degrees granted. Some of the US graduates may have been students from outside the US who received H-1B's or other visas and remained. We don't know how many of these US graduates entered the engineering job market – some of the bachelor's degree holders doubtless went into non-engineering jobs or moved to graduate studies in other fields, such as business administration, and some of the master's degree recipients went on to study for doctorates.

As of this writing the most recent available numbers on H-1B visas are from fiscal year 2019 (October 1, 2018 – September 30, 2019).[83] There were 138,927 H-1B petitions approved for initial employment and 272,175 for continuing employment, some of the latter having been received in the previous year. The total of petitions granted went from 332,358 to 388,403, for an increase of 16.9 percent. Of the total, 66.1 percent of approved petitions in fiscal year 2019 were for people in computer-related occupations.

The procedures that an employer has to go through to obtain an H-1B visa for an employee are complex. Nominally the services of an attorney are not required, but because of the rigor of the process there will almost always be an immigration attorney or at least a non-attorney specialist involved. The process incorporates several provisions intended to ensure that the visas do not unduly distort the labor market and thereby deprive US candidates of opportunities for which they might qualify. One such is that the employer must determine the local "prevailing wage" for the job to be filled. The US Government operates a Prevailing Wage Center for this purpose.[84] The process is supposed to ensure that employers don't use the H-1B program to undercut wages earned by American citizens and permanent residents, but the Government Accountability Office[85] (see also the full report)[86] found that "...over 50 percent of employers requesting H-1B workers between June 2009 and July 2010 categorized their prospective H-1B workers as receiving entry-level wages...". This is certainly not consistent with the notion that the country issues H-1B visas only to individuals of proven ability; doubtless many of the visa recipients have the potential to reach that status, but as entry-level employees they aren't there yet.

Another provision is the "labor certification"[87] through which the prospective employer attests that, among other things, the H-1B recipient will receive at least the same wages and benefits as the employer offers to other employees doing similar work, and that the H-1B recipient will work under conditions that are similar to those experienced by US citizens and permanent residents. Companies that hire large numbers of H-1B employees (15 percent or more for a company with 51 or more employees) may be subject to additional requirements; among other things they have to certify that they haven't "displaced" U.S. workers doing similar jobs within the past 90 days. These additional requirements apply only if the prospective H-1B holder is to be paid less than $60,000 per year and does not have a master's degree. The threshold of $60,000 has been in place since at least the year 2000 without adjustment for inflation.[88] This is well below current reports of the average starting salaries for engineers.

Only a few companies account for most H-1B visas. Here are the top 10, for 2018, each with its total number of visa applications (renewals and new, approved and denied):[89]

Cognizant Tech Solutions US Corp	13,584
Tata Consultancy Services Ltd	10,656
Infosys Limited	8,088
Deloitte Consulting LLP	6,372
CapGemini America, Inc.	4,912
Microsoft Corporation	4, 519
Amazon.Com Services Inc.	4,460
Wipro Limited	3,831
Accenture LLP	3,630
Apple, Inc.	3,123

Cognizant Technology Solutions Corp. is a US-based company that employs nearly 300,000 people of whom about 70% are in India.[90] Tata Consultancy Services Ltd. is an Indian IT services company, said to be the largest such company in the world, employing over 450,000 people.[91] Infosys, Ltd. is the second largest Indian IT company with over 239,000 employees.[92] Wipro Limited is a multinational provider of IT products and services, with over 181,000 employees.[93] So while the H-1B scene is dominated by IT services companies, both based in the US and offshore, American high-tech companies are also prominent on the list.

Because H-1B sponsorship is dependent on the employer, holders of the visas cannot change jobs as easily as can citizens and permanent residents, but the process is not complex. The new employer must submit a new petition for an H-1B visa, which is not subject to the cap in effect at the time. An H-1B holder who loses his or her job has up to 60 days (unless the visa expires before that) to find another job.

The annual H-1B cap is set by Congress. At present the cap is 65,000 regular visas plus 20,000 additional visas available to holder of advanced degrees from US institutions, plus an unlimited number of visas available to employees of institutions of higher learning, nonprofit research organizations, and government research organizations.[94]

H-1B lobbying

Lobbying is part of the process of maintaining the availability of H-1B visas. Lobbyists are required to report on their activities, creating an informative public record. In the House of Representatives, lobbying records are maintained by the Clerk.[95] One quick search, for lobbying records related to Microsoft and immigration for the first quarter of 2020, yielded Microsoft's report of spending $2,394,000 between January 1 and March 31, 2020.[96] In 2008 the Federation for American Immigration Reform (FAIR)[97] reviewed lobbying associated with three immigration-related bills before the House and Senate (not exclusively focused on H-1B visas) and attributed 13.9 percent of the lobbying expenses associated with these bills to the technology industry. Other types of lobbying organizations included general business, advocacy groups and educational interests, so the figure of 13.9 percent may not include lobbying funds provided indirectly by employers. The total spent on immigration-related lobbying during this time period was $345 million. While some lobbying organizations choose to break down their expenses by bill, they are not required to do so and others do not. For more detail see the full report, referenced above.

In addition to formal, reported lobbying, American business leaders have easy access to the media to promote their views. Microsoft chairman Bill Gates once called for an "infinite" number of H-1B's to be available[98] and Michael Bloomberg was reported to have also favored the end of the H-1B annual cap.[99]

H-1B abuse

Restrictions built into the H-1B visa system are supposed to help avoid abuse as the country brings in the best and the brightest from elsewhere to help build our economy. But it's not necessarily in the interests of employers to respect the *spirit* of the law as their lawyers help them to meet the *letter* of the law. (Immigration law is big business; the American Immigration Lawyers Association claims over 15,000 attorneys and law professors as members.)[100] There have been numerous publicly-reported disputes over the ways that employers have used H-1B's:

In October of 2014 about 250 employees at Walt Disney World were told that they were about to lose their jobs,[101] in many cases to employees of a "managed service provider"; in other words an Indian outsourcing company. Disney claimed that the jobs were lost through a "reorganization", and many of the displaced American workers were able to find other jobs within the corporation. Some workers, however, were required to help train the people who took over their jobs before they themselves were involuntarily terminated. Unless there's a union contract that provides otherwise, severance benefits are typically voluntary on the part of the employer. The employer can therefore use an offer of severance payments to motivate employees who are about to lose their jobs to stay long enough to transfer knowledge and skills to their lower-cost replacements. Disney called this a "stay bonus". Several former employees filed suit against Disney but then dropped the suit when it became clear that the law allowed Disney to do what it did.

While not reported as part of the Disney story, employees who refuse to train their replacements can be subject to termination for cause, which can in some jurisdictions make them ineligible for unemployment benefits. Severance agreements can also contain provisions that discourage former employees from going public with their complaints, although again this appears not to have been the case with Disney.

Also starting in 2014, Southern California Edison cut about 500 jobs in information technology, to be replaced by employees of Infosys and Tata Consultancy Services.[102] As with Disney, outgoing American employees were encouraged to train their foreign replacements. The US Department of Labor investigated the actions of one of these outsourcing companies and found no violation of law, stating that since the replacement employees were paid above the $60,000 annual salary limit, provisions in the law intended to provide additional protections to displaced America workers did not apply.[103]

Northeast Utilities (now Eversource Energy) laid off about 200 IT workers in 2014, replacing many or all of them with H-1B holders supplied by outsourcing services. A published report[104] provides the language that the employer used to discourage former employees from publicly discussing the situation.

I know of one similar case, involving another large corporation, that appears never to have been publicized. There are doubtless many others. Lower-level employees with mortgages and families, who can't afford to go without severance and unemployment compensation, have a lot to lose by going public in contravention of their severance agreements, whether they were replaced by foreign labor or terminated for some other reason. We may never know how prevalent these practices are.

Another form of abuse is "benching", whereby an H-1B visa holder remains in the United States, maintaining his or her visa, while waiting for a new job or assignment.[105]

Permanent Residency

Holders of H-1B visas may petition for permanent residency and thereby become legal immigrants. The process is complex and generally requires the assistance of an attorney. One requirement is that the visa holder's employer submit a permanent labor (PERM) certification establishing that no U.S. candidate is qualified and available to fill the position at its prevailing wage. Of course, the employer's human resources staff and immigration attorneys will make the strongest argument they can that this is the case. Sometimes they may go too far. Here's an example: In the waning days of the Trump administration, the Justice Department filed suit against Facebook (now Meta Platforms, Inc.), alleging discrimination against US workers in favor of current non-US employees then holding H-1B's. Facebook allegedly accomplished this through such methods as failing to advertise the positions on its own employment website.[106] The suit was settled in October of 2021.[107] Facebook denied any wrongdoing.

Conclusion

Since the H-1B visa and its predecessors have been in place for several generations, it's difficult to argue that its sole purpose has been to address temporary needs for specialized skills. And to be clear, individuals in possession of truly "extraordinary ability", at the peak rather than the beginning of their careers, can completely bypass the H-1B process, instead coming to the United States on O-1A visas[108].

We must conclude the obvious, that industry today uses H-1B visas not as a feedback mechanism to correct shortages in the technology labor market as they emerge, but as a *feedforward* mechanism to help employers to manage salary levels, and in general, maintain a pliant US labor force.

In fact, far from serving as a feedback mechanism, the H-1B visa prevents ordinary market feedback mechanisms from working. If the H-1B visa was truly part of a feedback mechanism then we would expect to see large numbers of H-1B visas issued in such fields as law and financial management, for the reason that the high pay levels in those fields reflect significant imbalances between supply and demand in favor of the latter. We might also expect to see some years in which the visa cap wasn't reached. It seems that as a society we trust the free market to provide adequate numbers of people trained in some fields of endeavor, but not in others. Or it may just be that the number of H-1B visas issued for an industry is inversely proportional to the political clout of that industry's workers.

The most persuasive argument in favor of issuing a large number of H-1B visas every year is that immigrant engineers start a disproportionate number of high-tech companies in the United States, thereby helping to create large numbers of jobs for American engineers and others.[109] Notwithstanding the record on this issue, we should consider the following:

Many immigrant entrepreneurs came to the United States as children so did not themselves participate in the H-1B or any similar program.

The government is aware that fraud and abuse occur in the H-1B process. In early in 2020 the US Citizenship and Immigration Service published a guideline on how to identify it.[110] Anonymous reports are accepted.

Some immigrant engineers do start companies that employ Americans, but many don't. Many appear to simply look for corporate jobs, doing the same thing that many American engineers do.

Immigrant engineers often enter the United States on student visas to attend graduate school. They typically don't owe large sums of money from their undergraduate educations, and are supported with grants, stipends and teaching assistantships, or by their families, while in US graduate schools. Many of them enter the world of work in better financial shape than their American peers and so are better positioned to forego the corporate world for life in a startup. In part, it may be that immigrant engineers are better represented among the founders of startups not because they're smarter than American engineers, but because they have greater financial flexibility.

As already discussed, there exist other mechanisms for allowing the immigration of foreign engineers with highly-specialized skills for which genuine shortages exist. We can, as we should, invite in these experts, who will necessarily be small in number, without distorting the American labor market.

Immigration proponents may argue that it's not possible to distinguish the extraordinary individuals referred to above, from engineers generally, especially since H-1B's are typically used to hire people at the beginning of their careers. Therefore, by this logic, to get the "stars", the American economy needs to absorb large numbers of immigrant engineers, likely enough to reduce salaries overall, notwithstanding the mechanisms that are supposed to prevent that. Let's accept for the sake of discussion that this argument has merit. American engineers can then argue with equal force of logic that since the presence of large numbers of foreign engineers in the labor market makes the occupation less attractive to Americans, there are *Americans* who would be studying engineering (and starting tech companies) but for being discouraged from entering or remaining in the profession by its generally limited economic prospects.

When immigrant engineers on H-1B visas do start US companies, they may outsource considerable work to their home countries, not necessarily or exclusively because specialized skills are more readily available in their home countries, but because doing so enables them to take advantage of both personal connections and lower cost. This practice can encourage the transfer to other countries of technology develped in the United States.

A significant number of H-1B applications are filed on behalf of individuals from countries that have adversarial relationships with the United States. Most prominent is, of course, China, which in fiscal year 2018 accounted for 11.2 percent of all H-1B petitions,[111] second only to India. In early 2019, Wei Sun, an engineer born in China but now a US citizen, was arrested by the FBI on suspicion of transferring sensitive information on missile defense back to China. In a plea deal on February 15, 2020, he pleaded guilty to one count of moving armaments (sensitive defense information) out of the United States.[112] In December 2019 Weiyun "Kelly" Huang was arrested on suspicion of setting up two companies that provided false employment documentation to Chinese individuals seeking H-1B visas. One of her "clients", Ji Chaoqin, was charged in 2018 for helping the Chinese government recruit spies from Americcan defense contractors.[113] Similar examples are routinely reported in the US press. In July of 2020, the US government ordered the Chinese consulate in Houston closed, claiming that a large number of Chinese postgraduate researchers had hidden their close relationships to the Chinese military.[114]

In today's global economy there may not be a sharp dividing line between products for commercial use and products for the military. That foreign nationals, especially from adversarial states, have access to non-public industrial and financial information is also a matter of concern. To be clear, though, I don't want to suggest that all H-1B holders from China or any other country represent a specific security threat to the United States; I'm sure that many Chinese citizens come here for the same reasons that others do, and are quite happy to not carry with them any residual loyalty to the governments controlling the countries they left. Charges are sometimes dropped before trial, and when a case goes to trial the result may not be a conviction. Plenty of native-born Americans have also acted against their country. Still, I believe that the employment of large numbers of people from generally unfriendly countries under the H-1B program as well as through other routes is less than ideal for the military security and economic well-being of the United States.

Replacing today's H-1B system

Is there any way to wean American industry off of the current H-1B system in a manner sufficiently kind and gentle so as to avoid excessive damage? "Cap and Trade" programs[115] have been used by the U.S. Environmental Protection Agency to reduce emissions from power plants. At the beginning of the program, sources were assigned "pollution credits" based on their historical performances. The availability of these credits was reduced for every subsequent year. A pollution source had the ability to build or install additional emissions-control measures, but these were often large investments, in the hundreds of millions of dollars. The source that chose to reduce its emissions then had unused credits that could be sold, effectively spreading the cost of the investment to other sources. Older plants extended their economic lifetimes by buying credits, or making more modest investments to reduce emissions. The cap-and-trade programs won support from the environmental community, which appreciated the certainly of its results, as well as from the power companies, which liked the way they gave each source the flexibility to deal with the rules in ways most appropriate to its individual situation.

A cap-and-trade program using H-1B credits could replace the current lottery. New H-1B's for most employers are already limited, in the aggregate, by Congress. What if the numerical limit on all H-1B's were reduced from year to year? There would need to be some formula for the initial allocation of credits, then allocations would be reduced for every subsequent year. Such a system would permit the US to reduce its dependence on foreign engineers while still allowing entry to particularly talented individuals who don't qualify for O-1 visas, and those who offer skills for which demand genuinely can't be met by the US labor pool. Employers in need of more credits than allocated can obtain additional ones through an open market, and those that need fewer than they have can use the excess to generate revenue. Thus the company that truly needs an individual trained in a rare specialty could get its worker through an H-1B visa albeit at a cost, while the company that uses the H-1B merely to reduce its labor cost might decide that economics now favor selling some credits instead and hiring in the domestic market. Corporations would no longer need to pay human-resources specialists and attorneys to prepare the most convincing applications possible; they will simply make their own private economic evaluations of their needs for H-1B personnel. Further, H-1B credit prices themselves would provide valuable information to policymakers.

A variant of cap-and-trade would be a government auction, such as has been used to assign portions of the radio-frequency spectrum to mobile telephone carriers. In this case the auction would be annual, with the number of credits reduced from year to year. The revenue from sales of rights to obtain H-1B visas could then be used to help retrain displaced engineers and train new ones.

Either cap-and-trade or an annual auction should be applied gradually. Business will need some time to adjust, as will the cohort of US-trained engineers.

Another alternative could be a requirement that US citizens and permanent residents be hired in some proportion to H-1B holders. Such a program would have to be carefully designed and managed, to prevent some obvious potential abuses: The job descriptions and levels would have to be kept similar to avoid a company's hiring an American accountant to balance a foreign engineer, and there would also have to be rules to prevent a company from keeping an American on the payroll for only enough time to justify the foreign engineer.

Employers wishing to hire foreign engineering employees under the H-1B visa program should be required to make an effort to recruit American engineers before being allowed to hire a non-American under an H-1B. An employer wishing to hire under an H-1B should be required to submit to the Department of Labor all applications received from American citizens and permanent residents, and be prepared to explain why those applications were rejected. The employer should also be required to advertise a verifiable prevailing salary, or not advertise a salary at all.

Finally, there should be changes to the rules governing labor certifications. In addition to justifying its claim that no US engineers are available to do the specified work at the prevailing wage, the employer should be required to review the labor market and determine whether any reasonable alternatives exist to hiring under an H-1B. One option could be to recruit an unemployed American engineer and retrain him to perform the tasks specified in the job description. Also, there should be a requirement that employers use a variety of common methods to seek job applicants before claiming that no qualified candidates are available – using services like LinkedIn, for example, that can target specific groups of American engineers likely to be interested in the advertised position.

The O-1 Visa

Unlike the H-1B, the O-1 visa requires that the applicant already be prominent in his or her field of endeavor. Evidence for this may include national awards, a record of publications, and even a history of commanding a high salary. There is an argument that the O-1 should be used more often, as a means of assuring that "the best and the brightest" of foreign engineers continue to be encouraged to work in this country, while minimizing distortion of the engineering labor market as a whole. Certainly an argument often made in favor of large numbers of H-1B visas, that granting fewer would deprive American society of the benefits of exceptional thinkers, is invalidated in part by the availability of O-1's. In 2019 the United States issued 31,831 O visas[116] of all kinds, including O-3 visas to dependents of individuals receiving O-1's.

L-1 Visas

In contrast to the better-known H-1B, intended for use by American employers who wish to hire foreign nationals, the L-1 visa is for use by multinational companies that wish to transfer foreign nationals to their operations in the United States. L-1A visas are granted to executives and managers, and L-1B visas to non-management specialists. Whereas H-1B visas are limited in number (with some exceptions) and recipients selected by lottery, there is no cap on L-1's so lotteries are unnecessary. The holder of an L-1 visa is only allowed to work for the sponsoring employer; transfer to another employer is not allowed. The Department of Labor is not involved in the processing of applications for L-1 visas.[117]

In 2019 the United States issued 157,708 L visas, which would include L-1A, L-1B and L-2 visas, the latter for dependents of the transferred employees.

The Dichotomy of Engineering

Information technology and engineering are so important to the United States that the U.S. Government sees fit to issue large numbers of visas every year to increase the national supply of information technologists and engineers. At the same time, there's a reasonable argument that engineering is relatively unimportant to US companies: The CEO who outsources large parts of her company's engineering work, and who lobbies Congress to authorize more H-1B's for work to be done here, is not likely to hand her company's accounting work to the lowest bidder, especially in light of the Sarbanes-Oxley Act of 2002, which imposes criminal penalties on executives of public companies who falsify financial reports. Nor is she likely to offshore her personal legal work.

Who speaks for us?

There is no question of who represents lawyers in the United States – the American Bar Association. Similarly, the American Medical Association represents the country's physicians. The existence of these groups still allows for physicians and attorneys to organize state-level societies like the California Medical Association, local societies like the Chicago Bar Association, societies devoted to practice specialties like the American College of Cardiology, and identity-based groups like the Hispanic National Bar Association. Notwithstanding the presence of all of these organizations, and their ability to represent their respective memberships in public discourse, all literate Americans know who speaks for our doctors and lawyers as a whole. Many also know that the American Dental Association speaks for dentists, the American Psychological Association speaks for psychologists, and so on.

Who speaks for America's engineers? No one entity, really. We have organized ourselves largely the way we identify ourselves – as chemical engineers, civil engineers and so forth. As pointed out earlier, we can identify ourselves simply as engineers but thereby invite confusion as to our role in society and level of professional training and credentials. It's possible for an organization that does not explicitly require professional licensure or even an engineering degree as qualification for full-membership status to name itself an engineering organization, as does the American Society of Plumbing Engineers.[118] We do have the National Society of Professional Engineers[119] (NSPE), but by its own choice of name it can't claim to represent the majority of working US engineers, who do not hold engineering licenses. Furthermore, with a membership of approximately 23,000 the NSPE is much smaller than the practiceoriented societies such as the American Society of Mechanical Engineers. In fact, there are so many engineering societies that many came together to form the American Association of Engineering Societies,[120] an organization that did good work but probably wouldn't have been be necessary if engineers themselves had ever created a credible umbrella society. The American Association of Engineering Societies announced its dissolution in 2020.

Unionization

The National Labor Relations Act[121] (NLRA) defines the rights of many workers to form unions, and the rights of their employers to oppose them. Some who are not covered by the NLRA are still able to form unions under other legislation. State laws may also affect union activities. Of course, the First Amendment to the U.S. Constitution guarantees the right of free association, but the NLRA and similar laws permit workers to create a specific kind of organization having particular powers. Mainly, this is the power of collective bargaining – the ability to negotiate with employers over such matters as compensation and working conditions.

The Bureau of Labor Statistics reports that for 2021, union members constituted 10.3 percent of the workforce, down from 20.1 percent in 1983, which was the first year for which these statistics are available.[122] Union membership among engineers is not the norm, although there are prominent exceptions. For example, engineers at Boeing are represented by the Society of Professional Engineering Employees in Aerospace (SPEEA).[123] That organization, in turn, is affiliated with the International Federation of Professional and Technical Engineers (IFPTE),[124] which claims 80,000 members in private industry and public employment. While the labor movement has historically tended toward positions associated with one major political party, the IFPTE has taken a bipartisan position on at least one issue, reforming visa programs.[125]

Engineers have been discussing the merits of unionization for decades. To illustrate the arguments, I bring to these pages a simulated debate:

Question 1

Is the union better able to promote an individual member's interests than the member can himself?

Argument for unionization: Yes, of course. An individual employee is just one person, has one voice; it is only through collective action that the employee's individual interests will be addressed, along with the interests of the employee's union brothers and sisters. The union contract will specify, for all to see, the relationship between employee and employer, minimizing capricious decisions of management. In the event of a dispute, the union should be able to promote the employee's interests using resources like experienced labor attorneys that the employee wouldn't be able to access so readily if acting as an individual.

Argument against unionization: No, not necessarily. A collective bargaining agreement will establish the details of the relationship between management and the employee, diminishing the ability of the employee to directly negotiate his own unique terms of employment. Rather, the interests of the membership as a whole will subsume those of any individual employee. For example, if most of the members of the union don't care about the imposition of a non-compete agreement, but a few care a great deal, the likelihood is that the union leadership will accept the non-compete agreements in return for some other concession that is deemed more important by more members, or by union leadership.

Question 2

Isn't union membership primarily for nonprofessional employees – technicians, members of the building trades, assembly-line workers, and so on?

Argument against unionization: Yes; the labor union movement, at least in the private sector, is largely oriented to the needs of blue-collar manufacturing and service workers.

Argument for unionization: No. The labor movement serves public employees of all categories and these workers are now the majority of union members. While union membership is today not the norm for private-sector engineers, many other workers with considerable education and in highly responsible positions do join unions. These include airline pilots, teachers, physicians, journalists, and even judges. There are many workplaces at which engineers are unionized, although more workplaces where they are not.

Question 3

Engineers already have a variety of organizations they can join – perhaps 20 national societies as well as specialized groups within these societies, regional groups and so on. More recently, these formal organizations have been supplemented with a variety of online communities. Can't these organizations collectively provide the engineering profession with a strong voice, so that unionization is not necessary?

Argument against unionization: Yes; well-established engineering societies have a long history of promoting the interests of their members, such as in Congressional testimony.

Argument for unionization: No. No professional organization can take the place of a union, since in practice these organizations are unable to operate independently and place their own members' interests paramount. Only a union – certified in accordance with federal law – can represent its members in a collective bargaining process. Unions can also actively pursue political goals on behalf of members since they are able to endorse candidates and sponsor political action committees (PACs).

Professional organizations are limited in other ways as well. They may have paid staffs but organizational officers are often volunteers. Most volunteer officers can't afford to take a year or more off to work without a salary, so typically they depend on their employers for continuation of their salaries and possibly to help pay expenses. Professional organizations also need employers to support members' travel to their conferences and trade shows, advertise in their media, and in many cases reimburse members' dues. These relationships discourage the organization from taking positions that favor employees over management.

Societies that serve an international membership may find it difficult to promote the interests of their US members versus those of members elsewhere. Societies that are dominated by academics and managers may view the interests of rank-and-file working engineers as secondary.

The largest engineering organization, the Institute of Electrical and Electronic Engineers (IEEE, operates worldwide and historically focused on dissemination of knowledge and setting of standards – not primarily on the economic interests of its membership. Concerns about the latter helped lead to a change in the organization's constitution in 1972 and eventually to the creation of IEEE-United States of America (IEEE-USA in 1998 to engage in "non-technical activities" on behalf of its members in this country.[126] But it's still not a union, and can't negotiate with employers on behalf of its members.

Question 4

In states that don't have right-to-work laws, unions may negotiate contracts that require membership as a prerequisite for employment. Then, they can set their own standards for membership, which can give union leadership the right to expel members for infractions that the leadership defines, including perhaps, public criticism of their positions. Isn't this a violation of free speech?

Argument for unionization: No. The union member would have voluntarily agreed to such a provision as a condition of joining the union. Therefore, it is not a violation of the member's free speech rights to allow union leadership, democratically elected, to set the requirements for membership. Similar provisions exist elsewhere, such as in non-disparagement agreements. For the union to be effective, it must have cohesion. Furthermore, if the absence of such a provision is important to its membership, members can vote not to include it in the contract.

Argument against unionization: The ability to maintain this level of control over members may be more important to the union's leadership than to its membership, so that leadership will tend to fight for it in negotiation even without the strong support of members. And since all labor contracts reflect compromises, the cost of rejecting this requirement can be high in terms of other benefits not received. Bottom line, membership in a union can inhibit free expression.

Question 5

By establishing an employment structure strongly influenced by seniority, as is common in collective-bargaining contracts, will unionization discourage the natural desire of engineers to improve themselves and maximize the contributions they make to their organizations?

Argument against unionization: Yes; union contracts tend to minimize individual performance differences and their attendant rewards as a way to maintain cohesion within the group.

Argument for unionization: No; an employee's status does not have to depend exclusively on seniority. A collective-bargaining agreement can give management discretion to promote deserving engineers, and may establish a bonus fund to reward exceptional performance. Further, while most engineers probably favor reasonable management discretion, limiting the exercise of this discretion via collective bargaining could help protect jobs held by Americans from being taken by lower-paid temporary foreign workers.

Question 6

At a time when American engineers often face competition from around the globe as well as large numbers of "temporary" individual competitors brought in from other countries, will the increased costs associated with unionization reduce the competitiveness of American employers and thereby reduce employment opportunities for American engineers?

Argument against unionization: That could happen. It's best to give American employers as much flexibility as possible to remain competitive.

Argument for unionization: It's insulting to assume that American engineers aren't astute enough to understand where their best interests lie. They will not allow their unions to take positions so aggressive as to threaten their employers' economic well-being. And history tells us that American employers who wish to offshore engineering work, or bring in foreign nationals to displace American engineers, do not require unionization of their US workforce as a prerequisite. An employer can do these things whether or not its US workforce is unionized.

Question 7

Could the rules imposed by a collective-bargaining agreement interfere with an employer's need to remain flexible?

Argument against unionization: Yes. Early-stage companies in particular need flexibility to respond to challenges as they emerge. Employees may sometimes need to work long hours to meet a schedule or solve a production problem. At times, people have to work outside of their job descriptions, but union jurisdictional rules that impose strict job responsibilities could, for example, prevent an engineer from fixing a machine.

Argument for unionization: If a company makes excessive demands on its employees, it should be prepared to fairly compensate them in some way – if not with cash, then stock options, time off, etc. A collective-bargaining agreement tailored to the needs of the workplace can give the employer the flexibility it needs to move quickly, while protecting employees from abuse. In the case of established companies with mixed unionized and non-union workforces, it's not unheard of for unionized technicians to make more money than non-unionized engineers when members of both groups are working long hours to meet a deadline.

Question 8

Does union membership present built-in conflicts of interest for engineers as professional employees?

Argument against unionization: Yes. The National Society of Professional Engineers discourages union membership in general, and in particular holds that "coercive tactics" such as going on strike are incompatible with the profession's responsibility to put the public welfare ahead of one's own economic status.[127] The Society also takes the position that engineers cannot identify as "labor" and at the same time claim that engineering is a profession.

Engineers who operate independent businesses that compete with each other risk violating the law if they act in concert to improve their economic conditions, e.g., cooperate in setting minimum hourly rates. Similarly, employed engineers should negotiate their working conditions individually, rather than collectively, even though collective action is lawful when properly engaged in.

Argument for unionization: No. Unions are comprised of "locals", which represent members working at one company or in one plant. Locals can represent the interests of their members on a granular level. The labor movement has recognized that representation of white-collar workers is an area of potential growth, and has been willing to make accommodations to the unique needs of these constituencies. In some cases, these accommodations have been specific to the occupations of relatively small groups of members. Conversely, negotiations over working conditions between management and individual engineers (i.e., not collectively) can lead to unproductive resentment among colleagues who believe that some are being treated better than others.

Members of other professions, notably lawyers, have formed unions and conducted strikes against their employers. For medical doctors, there is a United Salaried Physicians and Dentists Union. The AFL-CIO, an umbrella organization with which many individual unions are affiliated, has a Department for Professional Employees[128] to focus on their interests.

The ethical construct that engineers, especially those with licenses, are to put the interests of the public above all else is, in practice, an unreachable ideal. A literal reading would hold that engineers should seek only enough income to live at a subsistence level while perhaps paying off their educational loans; anything more would be to put the engineer's own interests first and thereby be unethical. I am aware of no profession that holds its members to such a standard. For example, financial advisors who accept the "fiduciary standard" that requires them to place their clients' interests above their own are not prevented from establishing a fee structure that they consider reasonable. The NSPE policy would allow self-employed engineers to set their own rates and policies while discouraging engineers employed by others from organizing to improve their own incomes and working conditions.

Conclusion

We engineers often think of ourselves more as management than labor. Even non-supervisory engineers make technical and purchasing decisions (albeit often subject to review), manage budgets, and take responsibility for the work of others in addition to their own. Our perception of, and desire for, independence does not encourage us to make common cause with our fellows by creating and joining unions. But it's not completely clear that corporate managers, especially those who are not themselves engineers, always think of engineers in the same way. Unionization is not an unreasonable response to this situation, but assuredly it isn't for everyone. Probably more engineers should be open to unionization, particularly in more established companies.

Self-employment and the the gig economy

Engineers often function outside of the traditional employment system, working as contractors or consultants. Both contractors and consultants, as defined here, are self-employed individuals, meaning that they engage with those who use their services as businesses rather than employees. As such, their compensation is reported by their client/employers on IRS Form 1099, which is used for transactions between businesses. Engineers can also be given similar titles, e.g., "consultant" by their employers but if their wages are reported to the IRS on Form W-2 then they are regular employees. Self-employed contractors may receive higher hourly pay than employees doing similar jobs, but they generally do not receive benefits given to regular employees, are not afforded certain legal protections, and are responsible for the employer portion of FICA taxes (for Social Security and Medicare) as well as the employee portion. They may be subject to state and local business taxes as well as various fees.

Sometimes an individual working as a contractor (or consultant) is formally an employee of a Professional Services Organization (PSO) or something similar. The PSO employee is assigned to a client organization and may work as a peer alongside the client company's regular employees. However the individual working for a PSO doesn't necessarily receive pay and other benefits identical to those received by the regular employees. The PSO serves as the "employer of record" and reports wages on a W-2. It may provide routine services like managing tax withholding, and it's easier for a client corporation to engage one PSO than work with dozens or hundreds of one-person businesses.

If you decide to go independent, talk to an accountant about whether to do so as a sole proprietor or through some other business entity. There are advantages and disadvantages to each choice, and differences between states, so you'll need some professional advice.

Is there a practical difference, then, between a contractor and a consultant? Often, a contractor, particularly in information technology, is expected to start with an input, say, a software functional specification, and generate an output, i.e., code corresponding to that input. Or a contractor working in testing will have an established protocol to follow. A consultant, on the other hand, is more likely to do work that is less defined, for example helping to decide on the best approach to solve a problem. Some companies consider a contributor with a commonplace skill set as a contractor, whereas an individual with more specialized skills, perhaps resulting from advanced education, and who brings in expertise that the company doesn't already have, is more likely to be seen as a consultant. It's more common to hear something like "we'll hire a dozen contractors to get the project done on time" than "we'll hire a dozen consultants to get the project done on time".

One good way to find out whether you're effectively a contractor or consultant is to think about who sets your rates – if it's the company you're working for then you're a contractor; if you set your own rates you're a consultant.

Normally, a contractor or a consultant is engaged by a client, directly or through a PSO, for a relatively short length of time – typically not more than a year – or to complete a specified task. Sometimes, though, these individuals work for a company for a much longer time, and become integrated with the employer's workforce to the extent that it becomes hard to tell who are the regular employees and who are nominally temporary. These are the "permatemps" – people whose functions and contributions are indistinguishable from those of regular employees but do not receive the benefits of regular employment. After a class-action lawsuit against Microsoft (Vizcaino vs. Microsoft)[129] was settled in 2000, thousands of permatemps were granted access to certain benefits, and the use of permatemps now appears to have declined considerably. Companies will now often require an absence of 100 days or so after an engagement, even if the contractor works for a third party such as a PSO, before the contractor can return to the client. This helps to counter any assertion that the contractor was in fact working for the client company.

Advantages and Disadvantages

An employer/client often has a financial incentive to classify workers as contractors or consultants rather than employees. Self-employed workers don't get benefits – no vacation time, employer contributions to medical insurance, etc. – and may be required to provide their own tools including computers. Workers employed by third parties receive benefits offered by their employers of record, which may be less generous than those provided to the client company's regular employees. Administration, e.g., payroll management, can be simpler for employers since there are no payroll deductions to manage. The burden shifts to the contractor or staffing company to manage withholding or make estimated tax payments as needed, deduct Social Security payments, arrange for insurance, and so on. Companies aren't taxed to fund unemployment insurance for contractors working under 1099's although contractors were made eligible for pandemic relief in 2020. Absent a contractual provision to the contrary, consultants and contractors can be terminated with neither notice nor severance pay, regardless of the company's policy toward its direct employees.

On the other hand, staffing a company with regular employees who receive reasonable pay and benefits is more likely to engender loyalty and focus. Toward the end of a project, employees probably be won't be putting as much effort into finding their next assignment as will contractors. In practice, it can be reasonable for a company to have a core of managers and senior employees, augmented at the top with consultants and on the bottom with contractors.

Legal situation

Since there is often an economic advantage to hiring a worker as a contractor rather than as an employee, the IRS tries to provide guidelines to help companies determine when it is proper to do so. The basics are given by the IRS in its Employer's Supplemental Tax Guide, publication 15A.[130] The determination isn't always clear. Quoting from the Guide: "In any employee-independent contractor determination, all information that provides evidence of the degree of control and the degree of independence must be considered." The employer is responsible for the classification of employees and contractors/consultants, and can be penalized for doing so incorrectly. State laws also have influence; California law AB 5,[131] which became effective on January 1 of 2020 requires the classification of certain contractors as employees. The law became well known for requiring ride-share drivers to be treated as employees, but it affects many others as well. It exempts "[anyone] who holds an active license from the State of California and is practicing one of the following recognized professions: lawyer, architect, engineer, private investigator, or accountant" from the broadest definition of an employee. Thus the plain-language meaning of the law appears to be that licensed engineers and their clients can ignore it; the law allows an employer to not classify a Professional Engineer as an employee. But what if the Professional Engineer takes a consulting assignment that otherwise would have been covered by the manufacturing exemption and thereby does not require a license? Would the Professional Engineer still be deemed to be "practicing"? In November of 2020, California voters approved a ballot amendment that exempts ride-share and delivery drivers from mandatory classification as employees, but other provisions of the law remain. No doubt AB 5 will be litigated energetically over the coming years, and its text itself may change. As questions like this are settled it's likely that other states will adapt similar – not necessarily identical – legislation.

Compensation management

Employers can pay their employees in a variety of ways, including base compensation, pay for performance (bonuses and commissions), mandatory payments to Social Security (separate amounts are paid by employees and employer), medical and dental benefits, paid time off, 401(k) matches and/or pension contributions, stock options, and discounts on such purchases as life insurance, parking and mass transit plans. Private-sector employers use knowledge of compensation conditions prevalent in their industries and geographic areas, and elaborate human resource systems, to determine how much of their revenue to allocate to employee compensation and how best to use these funds. They must keep their top management and star performers happy, their average performers reasonably content, and attract the new employees that they want. These processes normally proceed quietly, but on occasion do receive news coverage, such as when in 2018 Amazon.com announced a $15 per hour minimum wage for hourly workers, but at the same time eliminated bonuses and stock options.[132]

Practicing engineers are familiar with salary surveys; we are often asked to participate in surveys run by trade journals and professional associations, and there are websites like glassdoor.com that solicit and aggregate salary data across industries. But the surveys developed by compensation consultants and normally used for compensation management may be more accurate, as they don't depend on voluntary responses but instead use data supplied by the employers themselves. Human resource departments in companies of sufficient size may also have internal-only surveys, and government offers its National Compensation Survey (bls.gov/ncs). In any case, to set salaries, a participating employer will typically define a set of job categories, e.g., Senior Engineer, and then determine an average salary and range for each category. Range is usually expressed by percentile: Eighty percent of salaries will fall between the 10th and 90th percentiles. The employer then determines the percentile at which each Senior Engineer should be placed; presumably the strongest performers and more senior personnel within each job category end up closer to the tops of their respective ranges. The average for all Senior Engineers within the company or department will not necessarily correspond with the 50th percentile of the range; the company may feel that its generous stock option program, or its free lunches, or its high "coolness" factor may justify placing its personnel closer to the bottom of the salary scale. Conversely, a bias toward the top of the scale may be appropriate if the company wishes to fill a number of similar vacancies at once, or if the company feels that it needs to compensate employees for unusual aspects of the job such as long hours or frequent travel.

Over time, the salary bands change and presumably employees' salaries change with them. From one year to the next a company may keep an engineer at the same percentile, meaning that his salary increases with the overall market, or move his position within the salary range up or down. Of course, private-sector employers are not required to grant raises to employees who are not covered by collective bargaining contracts. Without raises, as the chart's range moves upward the employee's position in the salary chart moves over time toward the bottom.

Starting salaries in engineering tend to be high, compared to those commanded by four-year graduates in other fields. An engineer who gets a reasonable first job, and many do, may feel at the top of the world. Earning a full-time income is certainly different from being a student, and even if he took the time to get a master's degree, it's likely that his friends preparing for careers in other "professional" fields are still in medical school, law school and business school, or have delayed further education while they work for a few years. Once they graduate, his doctor friends are interns and residents, and his lawyer friends may be working as law clerks, while he, the engineer, may have advanced considerably in responsibility and salary. Over time, though, their career paths meet and then diverge – the engineer's salary grows at roughly the same rate, while the doctors move into private practice or other relatively lucrative areas, the lawyers go to work for prestigious law firms and start becoming partners, and MBA's continue their ascent up the corporate ladder having started a few rungs up. Doctors, lawyers and corporate managers can expect expect to earn incomes that increase over time, generally allowing them to maintain increasingly comfortable lifestyles throughout their careers. This isn't necessarily the case for engineers, who often experience salary compression.

An hour of engineering time has a certain value, determined in most cases by the employer. This value goes up quickly from the time an engineer starts working to the point where he has 5 to 10 years of experience, and then typically goes up more slowly after that. By the time an engineer reaches the age of 40 or so, the value of an hour of his time is not likely to increase dramatically from year to year; the increase in the engineer's value from, say, ages 40 to 50 is going to be much less than the difference in his value from 25 to 35. Thus, the age range from roughly 30 to 40 is when many engineers leave handson engineering for better-paid positions in sales or management, where values can rise again, earnings constraints are weaker, and opportunities for growth greater. The sad fact is that it can be financially advantageous for an exceptionally capable engineer to become a competent but less than outstanding middle manager. Also at some point, probably at or before age 40, opportunities to move into general management begin to diminish, as supervisory-level management slots come to be occupied by engineers who moved up earlier. Overall, as documented by David Deming of the Harvard Kennedy Center and the National Bureau of Economic Research in The New York Times, the substantial premium that STEM graduates initially earn over their liberal-arts peers is greatly reduced by age 40, as more of the latter move into management and other higher-paying areas.[133] This suggests that an engineer planning to stay in the profession should consider making financial commitments such as buying a house relatively early in his career, when his buying power as compared to his peers is highest. Considering college costs, it may also influence an engineer's decision as to when to have children.

As an aside, engineering societies, and publications directed to engineers, often survey their members and readers to ascertain their levels of career satisfaction. I have always found these surveys amusing. They purport to show that career satisfaction remains high as engineers progress through their careers. But they suffer from selection bias: Over time, the engineers least happy with what they're doing are the ones most likely to change fields and subsequently not be surveyed. A better approach would be to start with a cohort of newly-graduated engineers and then track them as they either do or don't continue in their engineering careers. The results might be different.

Why does this salary compression happen? Preceding the article in The Times, Deming and co-author Kadeem Noray published an academic study into its origins.[134] They were able to explain salary compression in STEM fields to be the result of rapid change in required skill sets. Other fields which don't change so quickly don't experience the severity of salary compression that engineers do, but in other fields salary compression is still related to the pace of technological change.

Engineering, and here I include computer science and information technology, is for good reason seen as exciting and dynamic. That's great, and helps explain why many of us entered these fields in the first place. But this dynamism means that the skills you brought to the workplace from college will turn stale over time. At first, as I have noted, you will likely find that your knowledge and skills are still current and that your effectiveness, as perceived by your employer, increases rapidly as you gain experience. But this happy situation is not likely to last forever. You will learn to navigate the workplace as well as you navigated college, but new graduates with the freshest, most in-demand skills will be entering the workplace right behind you. In other words, your value begins to plateau while others seen to offer more up-to-date skills begin to compete with you. Your employer may respond by limiting your wage increases while continuing to raise starting salaries to keep up with the market. Strictly speaking this isn't age discrimination, since there are other reasons the employer can cite for limiting raises for older employees, but for most of us it correlates closely. We engineers help set the pace for society's progress, then can see it leave us behind.

Types of age discrimination

The paragraphs above describe two distinct kinds of salary compression. The first applies to each engineer individually, wherein over time pay raises diminish and real earnings (accounting for inflation) flatten out, or perhaps even decline. The second kind applies between engineers, as starting salaries for new employees increase at around the same rate as the salaries paid to current employees, so that the premium paid for the older employees' additional experience is small. In extreme cases, there can be a salary inversion wherein the premium for experience is negative. We can call these "individual" and "inflationary" forms of salary compression.

There is a third kind. Salary compression may also apply to performance differences between engineers who are similarly situated. Within a given job band, e.g., Engineer II, there may be significant differences between individuals in their value to the company. Perhaps one engineer frequently takes continuing education courses, has conducted internal training for junior personnel, and is well known in his industry. Most would agree he should be paid more than the more typical Engineer II, who hasn't been so strong a contributor, but by how much? The higher-performing engineer's immediate manager probably has a fixed salary budget for her department. An extra dollar given to this outstanding contributor has to be taken away from the other engineers in the group, and the manager can't afford to have the average performers, who do still fully meet the employer's expectations, feel slighted and leave. Thus the premium paid to this engineer for his outstanding work is likely to be small, at least until he's promoted to Engineer III and moves into a higher salary range. We can call this "competitive" salary compression. This engineer in such a situation has an incentive to seek other opportunities where there might be a better match between his value and his compensation. In some companies bonuses are a tool to increase the pay differential among employees, but there may be limitations built into these systems as well – some bonus systems award extra compensation as a fixed fraction of base salary, so that the contribution of a bonus to increasing the pay differential between top and average employees will be small.

Salary budgets for groups of current employees are often, as noted, fixed, but requisitions to fill new or replacement positions may allow a wide salary range. So it's not uncommon for an engineer to find it beneficial to move every few years, once he's determined that his salary is rising more slowly than he thinks it should. It's notable further that there is a movement in the employment world, increasingly enforced by law, to not inquire as to the salary histories of prospective employees. Lawmakers impose these requirements from a belief that the effects of earlier discrimination will linger if an employee's salary in one job helps determine salary in the next. The effect of this trend on salary compression remains to be established, and whether it will become beneficial for engineers to move more often, has to the best of my knowledge not been studied.

A phenomenon related to individual salary compression can be called earnings compression. This occurs when employees are required to work extra hours, usually to meet a project deadline. Hourly employees will receive additional pay in accordance with the law, and if applicable under collective bargaining agreements. Generally, as "exempt employees" (exempt from wage and hour laws), engineers are not legally entitled to extra pay and often do not receive it. Under "time crunch" conditions it's not unheard of for the earnings of top hourly workers such as senior technicians to exceed those of engineers working on the same project, putting in the same long hours, and carrying greater responsibility.

While it may not be a legal distinction, engineers often perceive a difference between the occasional need to work extra hours and a schedule that as a matter of routine requires it. An employer may even, on occasion, accept a premium payment from a customer to expedite a project and require its engineers to work extra hours, but decline to pay overtime because doing so is not contractually or legally required. Many would consider this an abusive practice, and I have seen it happen. However, I heard of one situation where a customer company, aware of this possibility, used contract language to make sure that the supplier company shared the overtime premium with its employees.

A report on salary compression done in 2017 was published by compensation consultant Pearl Meyer and reported on by the The Society for Human Resource Management. Survey participants reported that the job category "Information Technology" was most commonly affected by salary compression with "Engineering/Science" coming in second. Salary compression in these two categories exceeded that in Finance, Human Resources, Sales, Manufacturing and Product Development.[135]

Are engineers ourselves complicit in the salary compression that the profession experiences? The American Society of Civil Engineers publishes a definition of engineering grades, from I to VIII.[136] Only Grade I, the most junior, includes no managerial responsibilities, and by Grade V the engineer is a supervisor. By the top, Grade VIII, the engineer is a high-level manager reporting to a board of directors or equivalent, but may still be consulting or doing highly specialized engineering work. Let's look at Grade IV, the highest grade at which the engineer has no responsibility for the supervisory management of others. That is, the engineer at Grade IV may have some responsibility for project management (a Grade V engineer could be a project manager) but is not a supervisory manager. That is, he is not a "boss" responsible for performance evaluations and so on. Grade IV corresponds to Project Engineer and also includes faculty at the level of Assistant Professor. The technical work done by a Grade IV engineer can include designing a complete project, so this is a position of serious responsibility. An engineer who enjoys the creativity of the profession and interaction with his peers, but doesn't want to be a project or supervisory manager, might want to stay in this grade for many years.

What is the minimum experience for Grade IV, according to the chart? Four years. But an engineer's working life could be forty years or more. We expect the engineer to continue learning and growing throughout these years and in fact his ability to remain licensed as a Professional Engineer in some states will be conditioned on his participating in continuing education to the requisite measure. (Grade IV is also the lowest level at which a professional license is required, according to the chart. Can this creative, engaged and licensed engineer, albeit perhaps not the most ambitious with respect to moving into management, reasonably expect to continue to receive good raises for, say, the next 35 years after attaining Grade IV? The answer suggested by experience is "probably not"; at some point, if he doesn't take on additional management responsibility he will stagnate, receiving nominal raises at best. The next level up is Project Manager and our engineer may just not want to do that. This, in a nutshell, is salary compression.

Every engineer, by age 30, should have at least the beginning of a plan to generate income from an alternative source. This can take many forms, including a significant investment portfolio, a part-time business, preparation for a career change e.g., enrollment in an MBA program, or having a well-paid spouse, preferably working in a different field for a different employer. By age 40, this additional income should begin flowing, and should be re-invested to the extent not needed to meet day-to-day expenses. Just be careful to avoid the risk of a common-mode failure. I wouldn't encourage you to start a landscaping business in an area where most of your potential customers work for the same company or in the same industry that you do. Ownership of a large block of stock in your employer also risks common-mode effects – if the business doesn't do well, both your job and the value of your stock may be affected.

Overqualification

Senior engineers often experience being denied employment because they're too good – too many degrees, too much experience, and too much knowledge, which in the employer's mind translates into wanting too much and possibly also an inclination to question corporate dogma. The employer has put some effort into quantifying the skill level needed to perform a particular job, and has set a salary range to match. Candidates offering qualifications beyond the typical make the employer nervous; she fears that the engineer will leave for another more suitable job as soon as one becomes available, or perhaps even that the engineer will become bored and inefficient because the job won't provide the intellectual stimulation that he might be seeking.

The designation of overqualification is not applied universally in the labor market. If you don't believe me, do Web searches on the terms 'overqualified attorney", "overqualified physician", "overqualified accountant", "overqualified dentist" and 'overqualified engineer" and note the number of results of each. Overqualification is a particular problem in engineering because engineering services are seen as commodities and professional growth opportunities are, in fact or by perception, not always available. Engineering work tends to go to the lowest bidder, in this case the person meeting the minimum stated qualifications, instead of the most capable and proven contributor. The law firm hiring a young attorney expects her qualifications to grow and has a hierarchy of legal positions that she can move into as that takes place, whereas in engineering there may be no such growth path available: The organization is "flat" meaning that there are few management slots available, and the employer may simply not be motivated to promote an engineer from one non-management job to another. Viewed this way, overqualification is seen as a form of self-fulfilling prophecy: Employers choose not to recognize the value of experience in their engineers, so that when experienced engineers do come through the virtual or physical door, the companies don't know what to do with them.

If the term "overqualified" is often applied as a pejorative to employment candidates, its near-synonym, the state of "under-utilization", is a fact of life for many engineers. In a typical medical practice, before you see a physician you have to at least register with a receptionist and may then see a less-senior health-care provider such as an assistant. To see a prominent lawyer, you may have to go through her office's general receptionist and then her personal assistant. Engineers, in contrast, have less support than ever; when personal computing became the norm, it replaced the levels of assistance (designers, drafters, technicians, department secretaries) that previous generations of engineers could rely on to improve their own productivity. Thanks to the technology that we in large part invented, engineers spend a good portion of their workdays performing tasks that don't really need their capabilities. I believe that this reflects the relative costs, i.e., the pay, of engineers as compared to these doctors and lawyers, and I conclude that as engineering salaries relative to other occupations have slipped, it has become easier for engineers to be seen as overqualified because the amount of time they are able to spend on professional work has in fact diminished. As an engineer, you may be well-qualified for engineering but highly overqualified for collecting and reporting travel expenses.

If you find yourself underutilized, ask for work that is more challenging, or make suggestions without being prompted to do so. Of course, don't make changes to your project without authorization – you may not know all of the reasons why it's being done in a certain way.

Under-utilization, salary compression and age discrimination all complement each other to help push productive engineers out of the profession.

Stock Options and Restricted Stock

Tax Warning: I advise you to meet with an accountant or tax attorney with expertise in this field before engaging in, or deciding not to engage in, any transactions involving your employer's stock options, the stocks you purchase with these options, or restricted stock. Under some conditions, you may owe tax on a transaction that produced no income for you. There are specific and complex tax rules that apply to options and restricted stock. As of this writing a mistake can expose you to a truly stunning level of tax liability, a situation that will probably never change.

Defined-benefit pensions for private-sector engineers have all but disappeared and aren't likely to return any time soon. Aside from starting one's own company, stock options are now the engineer's principal tool for accumulating significant wealth – enough to support a lifestyle of above-average comfort and a secure retirement – as a result of working in his principal occupation. A stock option is the right to purchase a share of a company at a specified price. Options are given to the recipient through a "grant". If the company is successful, the price of the underlying stock should go up and create an opportunity for a profit, in some cases a large profit. The offer of stock options has become a favored tool of compensation, especially for cash-poor startups. Even non-professional employees have become quite wealthy after picking the right startup and staying with it for a few years. Microsoft alone is estimated to have created about 10,000 millionaires by the year 2000.[137]

For many established corporations, stock options are traded as derivative securities, which can be bought and sold on exchanges. Incentive stock options (ISO's, also called qualified stock options) are a particular kind of stock option that can be issued to employees by a non-public company; that is, a company whose stock is not publicly traded. Small, technology-oriented companies in particular offer ISO's to ordinary employees as part of their compensation plans, but any qualified company can do it.

Restricted stock, as opposed to options, comprises actual shares of a company. The holder is prevented from selling restricted shares until certain conditions are met. Restricted stock as a form of employee compensation has gained popularity as a result of changes in tax law. Restricted stock *units* are a promise made by the company to an employee that it will grant the corresponding number of restricted shares at a specified time, and these are also seeing increased popularity. Restricted shares are inherently less risky for the employee than options because they don't require the employee to make an up-front investment; however, restricted shares may be taxable according to their value when received.[138] Get advice from someone who understands the applicable rules in detail.

But for all of the wonderful stories out there about humble employees becoming rich beyond their dreams (some of which you are likely to hear during an interview with a startup), stock options, in my view, are routinely oversold.

How are they oversold? The first requirement for a startup's stock option to benefit you is that the company has to do well and then undergo a "liquidity event", typically the company's sale to another company, or to the public via an Initial Public Offering (IPO). All of the risk is on you; an option in a failed company is less valuable than the stock certificates repurposed as wallpaper during the Great Depression. But there are other reasons as well why you might not get much out of your option grant.

You have to be at the company long enough for the options to vest. A vested stock option is actually yours; you can't buy stock with an option that hasn't yet vested. (A restricted stock unit similarly does not become a share of stock until it vests.) Your offer letter may have committed the employer to giving you some number of stock options, but the options probably won't be available immediately; they may vest over several years, commonly starting a year after the option grant. A typical schedule would have 20% of the options vest one year after the grant, the next 20% vest after the end of the second year, and so on. Once an option vests, you are able to exercise it and buy a share at the specified price. Two thousand vested options let you buy two thousand shares.

If you wish to exercise options and then hold on to the shares, you'll need the cash to do so. Under some circumstances, you may be able to sell the options without having to invest any money. You should know whether there is a time limit for exercising vested options even if you remain employed at the company.

You also have to be aware of what will happen if the company or its assets is sold. Vesting may accelerate if the company undergoes a liquidity event before all options have been vested.

Find out what will happen if you leave the company, both to vested and non-vested options. If vested options are subject to forfeiture when you leave, consider whether to exercise them before you lose the ability to do so. There may be favorable treatment for non-vested options, e.g., accelerated vesting, if you leave involuntarily rather than voluntarily.

All told, it would be a good idea to sit with a knowledgeable accountant or attorney for an hour or so and review the options grant as soon as you receive it.

You may lose control of your ability to sell the stock under favorable conditions. The payoff from your stock options comes as a result of the company's liquidity event (for a startup) or an increase in its share value. In the case of an IPO, there will generally be one or more underwriters, which are the investment banks that lead the process of taking the company public. It is in the underwriters' interest for the newly-issued IPO stock to remain high in price in the days and weeks following the company's going public. To help achieve this they may require the company to prohibit employees from selling their personal shares for some time after the IPO. By the time you're allowed to sell the stock, it could be worth considerably less than at peak. Even after that, your employer may decide to issue additional rules concerning this stock, to avoid the possibility or perception of your trading on insider knowledge even if you have no access to the company's unreported financial information. Restrictions like these may be instituted after the IPO; they won't necessarily be spelled out in your option grant.

If your company is sold or merges into another, there may be a stock component of the sale price; you would receive shares of the new company in return for your shares of the old company (or options to buy the old company's shares) according to some ratio. Would you have been attracted to this new company's stock on the open market? Consider whether you want a large fraction of your net worth tied up in the stock of that one company. Make sure you understand any restrictions on what you can do with the new company's stock.

Your ownership share will likely be reduced over time. Your options will be for a specified number of shares at a specified price. You may be able to calculate your prospective percentage of ownership, but there is no guarantee that it will be maintained. As the corporation acquires additional equity financing in return for issuing additional stock, your options do not adjust; your ownership will be diluted. Other stockholders, such as angel investors and venture capitalists, may be treated better. A board of directors may issue more than one class of stock, each with differing rights, and it's likely that shares corresponding to the grants issued to employees under ISO's will have the lowest level of such rights.

What are the rights that these favored stockholders may receive? First, ISO's are generally for common stock, which is what people usually mean when they talk about stock. The best-known stock indexes refer to prices of common stock. Early-stage financial (that is, non-employee) investors may get something else, preferred shares. Preferred shares put their holders second in line after creditors in the case of corporate liquidation. With small startup companies, liquidation is certainly a possibility, and the company's assets may have some value. Only after the holders of preferred shares receive their due from company assets, will holders of common stock divide up what, if anything, remains. Even the founders of a company can find themselves getting nothing from a liquidation while preferred shareholders get everything that's left.

Preferred shares may have additional special rights. While preferred shares generally are non-voting, a majority of preferred shares may be required to approve certain corporate decisions, regardless of the proportion of preferred shares to the whole. Preferred shares may have special rights to vote for directors. This may be implemented as an agreement with investors that preferred shares will be able to elect a specified number of directors, regardless of their fraction of ownership. Preferred shareholders may also have the right to receive certain information on the company's activities, that need not be disclosed to holders of common stock.

There can also be different classes of common stock affording different rights. Financial investors may be given shares with drag-along rights, which allow the holders of a majority of such shares to compel the sale of the company even if they don't represent the majority of company ownership. The reason is that a company can be more attractive as an acquisition candidate if a hundred percent of it can be made available for sale. The converse of drag-along rights are tag-along rights, which guarantee that minority shareholders can participate in any sale; without these rights minority owners can be denied participation.

When the company issues new stock, financial investors' shares may be adjusted in quantity or value to keep the original investors "whole;" that is, to maintain their percentage of ownership. This is called dilution protection, or sometimes anti-dilution protection. Employees may thereby have their options diluted twice in one round of equity financing – first by the original issuance of stock, and second by the additional stock issued to some earlier investors to maintain their percentage of ownership.

For more information about incentive stock options, see Investopedia.[139]

Restricted stock carries many of the same limitations as options, including vesting over a length of time.

Options can be wonderful things. However, given the risks and restrictions involved, an engineer shouldn't place excessive value on them. The kind of option offer that non-management employees are typically given is not often, in my opinion, adequate to turn an unattractive offer into an attractive one if there are credible alternatives. The younger and smaller the company, the more reliance it is likely to place on options as its primary currency of compensation, but then, the younger and smaller the company the more likely it is to never fulfill the options' implicit promise. An article in the New York Times Magazine sums up the situation in one sentence: "Startups explicitly use a lottery system, known as stock options, to entice young people to work for nothing."[140] Your tolerance for a such a situation may be inversely proportional to the amount of money remaining on your student loans. That said, startups sometimes don't ask people to work for "nothing", offering salaries that are competitive even before adding in options or restricted stock.

The best advice I can give you is to do what good investors do; that is, try to make your decisions based on logic and keep emotion out of the decision. You can become wealthy through options and restricted shares, but be financially and mentally prepared for the value of your options or stock to go to 0. Both good and bad outcomes have happened and will certainly happen again.

The Human Resources Department

Every company has some type of Human Resources (HR) function. In the smallest of companies it's embodied in the sole proprietor. In slightly larger companies someone with a title like executive assistant or office manager has HR as one of her roles, sometimes with the assistance of a part-time consultant, and in companies of medium size and above Human Resources can be a whole department.

These functions used to be called "personnel." The term "human resources" is supposed to represent the resources and benefits, such as insurance and educational programs, that the employer through its HR function makes available to employees. Indeed, larger departments have specialists who work on nothing else. "Resources" also includes the expertise necessary to comply with government mandates applicable to various aspects of employment. But for me the term also carries an implication that the human is the resource. For management, Human Resources is the function that secures the people necessary to meet its needs (and then removes them when they are no longer required). Some resources, like oil fields, are used up and then abandoned. I'm sure that this is not the sense that corporate use of the term is intended to convey, but some employees may develop such a view over time. The HR function can also carry a trendy name like Human Capital, and the head of HR can be the Chief People Officer.

The first thing to know about the HR function is that it is part of management and not intrinsically your advocate. Its relationship to you is situational and transactional. If you happen to be the person that management asked HR to recruit in order to meet a pressing need, HR will certainly be solicitous. If you need help straightening out a compensation issue or pursuing a claim against employer-provided insurance, you should expect their determined support. HR may be able to help you to advance in your career by offering in-house training and helping to make arrangements with outside training providers. But if management determines that your services are no longer necessary, the process of "decruiting" you may not be particularly enjoyable for the HR staffer assigned to execute it, but it's equally part of her job. If your employer is small enough, you may even have the experience of being walked out the door by the same person who earlier greeted you at the door as a new employee.

A lot of HR time these days seems to be spent on adjudicating claims relating to discrimination and harassment, and the natural inclination of HR will be to protect the organization first. An example, we hope not a typical one, is offered by software engineer Susan Fowler[141] who recounts her travails at a well-known technology company.

Human Resources specialists consider themselves professionals and are represented by the Society for Human Resources Management (shrm.org). The SHRM has established a Code of Ethics[142] to govern the activities of HR professionals. However, to me it reads more as a general declaration of principles rather than a code that prescribes and proscribes behaviors that, respectively, are and are not desired. For example, there is no explicit requirement that HR notify job candidates when they are not picked. Further, it is self-referencing, demanding ethical behavior without defining what the code's writers believe that is. The SHRM has disciplinary procedures to address unethical conduct, but since HR practitioners don't need licenses or to maintain membership in the organization, the practical effects of such discipline are not clear. While disciplinary actions against holders of licenses to practice (as engineers, physicians and so on) are generally public documents, there appears to be no requirement that disciplinary procedures against HR practitioners be made public. Finally, in dealing with HR you're unlikely to know who, as a member of SHRM, has voluntarily subscribed to its ethical code, whereas in professions for which licenses are mandatory to practice you do have an assurance that a code of ethics applies.

All of this means that your dealings with HR should always be respectful, even friendly if appropriate, but keep in mind that you may not be their first priority. This goes as well for the independent recruiters who work with HR. You may, for example, see several ads posted by different recruitment agencies for the same job because HR engaged several competing agencies to recruit for that position. Many consider it an abusive practice when HR does not disclose to these agencies that it is using more than one of them for the same job. The presence of different ads for what is in fact the same job may confuse candidates trying to assess the state of the labor market. A company itself may advertise nonexistent jobs to create a file of potential future candidates, or as a way to improve its public image since such advertising implies growth and general corporate health. Once I was called to an interview only to be told by the (non-) hiring manager that the position had already been offered to another candidate. The manager was still, though, curious about me and wanted to meet me. He said, "I asked [the HR person] to explain to you that we had filled the position. Didn't she tell you?" No.

Acquaintances who work in HR have told me that they are uncomfortable doing things like this, but that they have no choice since it's what management requires. Perhaps a more authoritative code of ethics would then help them to resist unreasonable demands.

The employment agreement

The employment agreement is a contract between the employee and the employer and is generally executed at the beginning of employment.

The wording of the agreement reflects and enshrines the disparity in power between employer and employee. In theory the terms of an employment agreement may be negotiable, but reality may well be different. Lawyers use the term "contract of adhesion" to refer to a contract that, almost literally, you're stuck with. Talk to your lawyer about how this designation could affect the way your agreement is interpreted by a court.

The employment agreement obligates the employee to abide by a set of restrictions dictated by the employer. It generally obligates the employer to do nothing, or nothing that isn't already required by law. Remember that the agreement was prepared, almost certainly, by a lawyer, to advance the employer's interest to the extent possible while still being acceptable to the employee. Absent the client's direction to the contrary, an agreement that doesn't aggressively promote the employer's interest would, arguably, be an indication of legal malpractice. An agreement offered by an employer that dominates its industry or technology, may reflect *monopsony*, a condition wherein one purchasing entity so dominates the market for some good or service (e.g., specialized labor) that it can unilaterally set its terms and price. It's the buyer-side equivalent of a *monopoly*. When it's a small group of employers that collectively dominate a segment of the labor market, they can be said to constitute an *oligopsony*.

A draft of the employment agreement should be sent along with the formal offer letter to the successful candidate for employment. We engineers often become impatient with legal verbiage; who reads the 20-page license "agreements" that come with software products? That the employment agreement is non-negotiable is clearly implied. Nevertheless it can be a mistake to sign one as presented, before at least understanding it fully and asking questions as necessary. In responding to a proffered employment agreement, you may need both a business and legal strategy. I leave advice on the latter to your attorney, who should review the document and explain it to you, but I offer the opinion that your business strategy for career management should drive your legal strategy, and not the other way around. In other words:

Do *read through the agreement* until you understand all of its business and career implications, referring any ambiguities to your lawyer as needed. Remember that some of the terms in the agreement, even if they are already familiar to you, may have different legal meanings than the meanings they hold for you in common use.

Do *make note* of any provisions that would impose excessive restrictions on your personal or professional life, for example an assignment provision that would force you to turn over your Great American Novel to the employer. I once signed an agreement that, among other things, gave my employer the rights to any works of art I produced during my term of employment. I will confess here, publicly, for the first time, that I failed to submit my vacation photos to corporate management.

Do *engage in a "what-if" exercise* with respect to the agreement's provisions. You can do this on your own initially, and then bring any questions to your attorney. "What if" you have to move to a new city for the job, only to see the office in that city close after you've been there for only six months? Will the employer pay to move you back? Your attorney may be able to suggest other realistic, if unusual, situations that you wouldn't have thought of.

Do not immediately return your lawyer's marked-up copy of the agreement to the employer if you see the need for it to be changed. What you may want to do is keep the discussion between principals, rather than between lawyers, for as long as possible. Your lawyer should be in the picture, but only in the background at this point. You and the employer should be discussing, in business terms, your objections to any business implications of the agreement. Once you've reached an understanding as to how the intent of the agreement should differ from the standard wording, it's time for your lawyer to work with the employer's lawyer to change the wording of the standard contract to reflect this new understanding. Of course, if your lawyer has technical objections to the language of the agreement as proposed, notwithstanding its business implications these should be brought up as well.

If you do object to the wording of the agreement, the employer may express puzzlement, telling you that everybody accepts the same agreement. First, unless the manager or human-resources representative telling you this is very high in the organization, she may not know what the company's most senior personnel have signed. Second, even if the claim is true it's not fully relevant, inasmuch as some provisions affect few employees as much as they do engineers. Third, high-level executives may be subject to agreements similar to or perhaps in ways even stricter than the one you're being asked to sign, but they may also have negotiated additional benefits including protections against involuntary termination resulting from a change of control.

I have never heard of a lawyer being disciplined for inserting a requirement into a contract that she knows may not be enforced. The term for this is *in terrorem* and as the Latin suggests it is used as a means of intimidation. In other words, the employer may be able to take advantage of your lack of familiarity with applicable laws and court decisions, i.e., your not having an astute attorney, to induce you to behave in ways that unnecessarily disadvantage you. A ruling by the Supreme Appeals Court of West Virginia framed the issue clearly: "... where savage covenants are included in employment contracts so that their over-breadth operates, by in terrorem effect, to subjugate employees unaware of the tentative nature of such a covenant, we will find the covenant void."[143] (Laws and their interpretations change all the time, and can be different in different places, so as always you must discuss with your lawyer how this principle might apply to you.)

Your employer can present new or modified terms to you at any time, so don't assume that once you've started working for a company under an initial agreement, the negotiation is over. Years into a job you may receive a demand for a new concession, especially if your employer is in the process of being acquired. This can work the other way as well. It may come to pass that you wish to renegotiate an existing agreement to establish terms more favorable to yourself. You may want to renegotiate if, for example, the employer materially changes the nature of your employment by requiring that you spend half of your time traveling when this was not initially part of the job description.

Don't try this on your own. You really do need a lawyer on your side as much as the employer has one (or a whole department of them) on its side. Most of the time an engineer will have only limited negotiating power with a prospective employer. But if what the employer demands is onerous enough, it's good to be able to walk away and know exactly why you're doing so.

Some typical provisions of employment agreements follow:

At-will employment

Ask your lawyer to explain what kind of employee you are and how local laws apply. Depending on applicable law and whether you are covered by a collective-bargaining contract, the agreement may remind you that you are an *at-will* employee, meaning that you can be terminated "for any reason, or for no reason". However you can't be terminated for the *wrong* reason, i.e., an illegal one, such as on the basis of your ethnicity. Most U.S. employees are at-will, even if they don't have employment agreements that make this explicit. If you're working under a consulting contract then that contract's provisions will apply. Employees working under *for-cause* rules rather than at-will rules cannot be terminated at will, but the employer's poor financial situation, whether or not you had anything to do with creating it, may constitute adequate cause. At-will employment is reciprocal, of course. You are free to leave whenever you want. However, you'll probably be carrying with you some significant restrictions, as set forth in other parts of the employment agreement, and the agreement may require you to give a specified term of notice before you leave voluntarily. Whether you're still an at-will employee if you're required to give specified notice is a question I'll leave to a lawyer.

If your employer terminates you, you may be eligible for a term of unemployment compensation, which is insurance provided by government and funded, at least in part, by taxes on employers. The details are set by state laws and vary. In general, if you are terminated for cause, e.g., for misconduct, you are either not eligible at all for unemployment compensation or your payments are subject to delay. In some places, certain conduct outside of the workplace such as offensive social media posts can be cited as cause. If you are terminated without cause, you are probably eligible for compensation. Generally, if you just quit, you're not eligible, but your state may offer certain exceptions such as if the employer required you to work under unsafe conditions.

If you're involuntarily terminated, your Human Resources department should give you the documentation you need to apply for unemployment compensation and may be able to assist you in doing so. Become familiar with your state's rules, which should be readily available online. It is not common for a terminated worker to engage an attorney, but that option may be of value in some cases.

Don't relocate your life to your office, no matter how much you like your job or how long you've been doing it. You shouldn't have so much personal property in your workplace that you can't easily carry it out if suddenly told to "clear out your desk" as part of a termination. This might especially be a problem for those who commute by public transportation or bicycle. And remember that computer equipment made available to you by the employer remains the employer's property, with content and transactions accessible to the employer.

The assignment clause

The assignment clause, at a minimum, transfers to the employer ownership rights in business-related inventions and other work products made on the employer's time during the course of employment. Many assignment clauses do much more, requiring the employee to turn over all intellectual property developed during the course of employment regardless of whether it was done on the employer's time, made use of the employer's facilities, or is at all related to the employer's business. The definition of covered intellectual property may be quite broad, extending all the way to "works of art"; conceivably, your employer can claim the doodles you make while you're talking on the phone. Why they would ever want to do that is a good question, but such a provision could be applied, for example, to a serious part-time artist or songwriter. The employer may claim the rights to intellectual property created after employment, if the employer believes that the intellectual property derives in some way from your work for that employer. The employer may even put in a requirement that you make yourself available after your employment ends, to help defend a patent that you were granted as a result of your work there. Such a provision has at least the potential to interfere with your ability to work for a future employer.

A provision that gives the employer ownership of work products related to the employer's business may not be restricted to the work that the engineer is doing; conceivably a large employer could claim the rights to inventions relating to lines of business that the employee isn't even aware of.

The engineer is likely to find that assignment clauses are imposed with particular enthusiasm by early-stage companies, whose investors are concerned that there be no impediment to selling the company to another or through an IPO.

George Laurer helped invent the linear bar code in the early 1970's while an electrical engineer at IBM. His obituary in the Wall Street Journal[144] quotes him as saying "I didn't get a dime" from the invention. Our professors often do better than that. Unlike engineers in industry who are generally required to give their patent rights to to their employers with at best a nominal payment as determined by the employer, professors may be able to derive significant income from their patented inventions. Here's the formula in effect for faculty at Yale, as of the time of this writing:[145]

> For the first $100,000 of net royalties: 50% to the inventor(s); 50% to support University research

> For the next $100,000 of net royalties, $40% to the inventor(s), 60% to support University research

> For net royalties above $200,000, 30% to the inventor(s), 70% to support University research

See the referenced document for the definition of "net royalties".

An invention by an engineer employed in industry is likely to be used internally, and it can be more difficult to determine its financial value than when it is licensed to third parties. Still, I have seen companies spend a lot of money both defending and challenging patents, even when no particular reward goes to the inventors. I heard of one company that holds a lunch to honor each patent recipient.

How likely is it that an employer would actually invoke the assignment clause in some matter unrelated to the engineer's services to the company? I've never heard of it happening, and some states may prohibit it, but if, for example, you were to create the next "Harry Potter" while being under an assignment clause that extended to unrelated work, you might well hear from your employer's lawyers. A broad assignment clause could also be used to suppress your freespeech rights: An employer (past or present) could argue that it owns any statement of yours that criticizes it since such a statement is clearly work-related, and so the assignment clause gives it the right to direct you not to publish the statement. There may also be a different clause that has a similar effect, namely:

The non-disparagement clause

Related in ways to the assignment clause, a non-disparagement clause in your agreement will prevent you from saying bad things about your employer. Typically, it will have been written broadly, to cover what you say in any medium (verbal, electronic, written), in any form (assertion or opinion), in any venue (over drinks, in an employment interview with another company, on social media, in a newspaper article) and for all time. Ask your lawyer whether the employer's definition of disparagement extends to assertions that are true.

As a practical matter, some infractions are going to be more important than others. Your pillow-talk conversations are unlikely to come to the attention of the employer, while a post on social media can go around the world in seconds, and the employer may be monitoring public posts for mentions both good and bad.

Disparagement clauses are usually not mutual, in that they do not prohibit the employer from saying bad things about you. However most employers won't disparage an employee or former employee anyway, having been instructed by Human Resources simply to confirm the dates of employment if asked.

A non-disparagement clause may be incorporated in an initial employment agreement, but it is more commonly part of an involuntary separation agreement. Here, the (about to become former) employee agrees to a set of provisions imposed by the employer, in return for a severance package which may include a term of salary continuation, insurance continuation, and so on. In most cases the employer has no obligation to provide such compensation after employment ends, so the package can be made conditional on the employee accepting these restrictions.

It will be good to consult with a lawyer before signing any separation agreement, including one with a non-disparagement provision. For example, if you were required to train foreign nationals to do your job before you lost it, ask your lawyer whether you would be able to file a discrimination case (because you're an American citizen or permanent resident) even if you accede to the separation agreement. Your employer should give you a reasonable time to have the agreement reviewed. A refusal to sign will cost you money, but you may decide it's worth it to forego the company's severance package and be able to go public with your story.

The non-compete clause

A non-compete agreement is a commitment on the part of the employee not to compete with his employer. Of all the provisions in an employment agreement, this is the one that's probably the most contentious. Non-compete agreements naturally prohibit the employee from competing with the employer while working for that employer, but here our concern is with the employee's obligations after termination. In its usual form in the United States the non-compete imposes no obligation on the employer, but in some countries and at least one US state there is a practice of "garden leave" whereby an employee remains on the payroll while not permitted to work for others. The wording of the non-compete provision generally specifies the activities defined as competitive (which may be anything related to what the employer does, whether or not you've been involved in it), the geographical area in which the employee is restricted from these activities (it could be the whole world), and the length of time that this restriction will remain in force (rarely more than a year for non-management employees).

The typical non-compete will restrict the employee's ability to compete with the employer in any capacity (i.e., as employee, investor and so forth) for the specified period of time following termination. Despite industry's occasional claim that non-competes exist primarily to discourage employees from leaving, I have never seen a non-compete that doesn't also apply in cases of involuntary termination. Another oft-cited justification is the need to recover an employer's investment in training the employee, but the employer rarely commits to a specified amount of training, and typically will not have participated in paying for the employee's undergraduate education.

In the past, non-compete agreements applied only to the most senior personnel in an organization, but this practice is long obsolete. A research study by Evan P. Starr and others in 2014[146] found that almost 20% of all US labor force participants were covered by non-competes. The occupational field with the greatest proportion of non-competes was "architecture and engineering", followed closely by "computer and mathematical" jobs.

Non-competes typically make no reference to the length of prior employment, so that it's possible in principle for an individual to be employed by a company for a week and then be barred from related work for a year or perhaps longer. This is just one of the nasty surprises that could await someone who signs a non-compete without understanding it.

Whether a court would enforce a non-compete under these conditions is another matter, but it could be expensive to find out. In reality, courts will sometimes "blue-pencil" a non-compete, allowing some provisions to stand while striking down others. This practice, sometimes also called reformation, has been criticized for encouraging employers to demand everything they can possibly think of. It's then up to the engineer to challenge overly-broad provisions at his own expense. Even if one worker's agreement is blue-penciled, the struck-down wording may remain in agreements presented to new employees. These new employees will in all likelihood just sign the agreements and do their best to comply with provisions that could prove unenforceable if tested. Current employees will rarely if ever be made aware that some provisions of their own identical non-competes have been invalidated. In many cases, then, compliance will comprise the employee's surrender to intimidation and an unnecessary gift to the employer. The research study referred to above even found non-competes used in states where they are by statute unenforceable. See the previous discussion of *in terrorem* provisions in an agreement.

While the wording of a non-compete provision may be the same for everyone in a company, its consequences can be very different. Some jobs are by nature highly mobile, by which I mean that the skills they require for success can easily be utilized in other venues, while the skill sets used in other jobs are not so mobile and can be of similar value in only a limited number of other places. The holders of the latter class of jobs, many of whom are engineers, are the people most seriously affected by non-compete clauses.

Let's start at the top: Consider as an example the career of Lou Gerstner. Mr. Gerstner was educated as an engineer at Dartmouth and then got a Harvard MBA. After working at the consultant McKinsey & Company, he became President of American Express, Chairman and CEO of RJR-Nabisco, and then CEO of IBM. It's generally accepted in business, and it certainly was by the various Boards of Directors that hired Mr. Gerstner, that CEO-level skills in organizing companies, dealing with investors and so on, are readily transferable between companies and even industries. In other words, these skills are portable. It's hard to imagine two large companies more different than RJR-Nabisco and IBM. The employment agreement that Mr. Gerstner presumably had with RJR-Nabisco would not have contained a non-compete provision so broad as to restrict him from moving to IBM.

Portability is high for many other corporate jobs as well. Finance and accounting skills are generally transferable across a wide range of companies and industries. Many marketing jobs are readily transferable; if you can organize participation in a trade show for a company in one industry, you can probably do so for a company in another. Even many software-related skills are transferable; if you're employed by a company to design its website or manage its e-mail services your skills can be equally valuable to an employer in a different industry. For people in any of these jobs, a non-compete, however nominally strict and tight, should not unduly restrict their ability to find equivalent or better employment when they wish to seek it.

Now consider the case of a senior-level engineer. To reach this level, he must have demonstrated specialized, in-depth knowledge of his field. This knowledge is likely to be of greatest interest to companies in the same industry; that is, his current employer's competitors and possibly his employer's customers and suppliers. The senior engineer who is, say, an expert in design of aircraft propellers has at best a limited set of prospective employers who might be willing to pay him an appropriate salary for his expertise. If he is subject to a non-compete agreement, for the term of that agreement he is restricted from working for most of the companies that might otherwise consider hiring him. Thus the senior engineer looking for a job starts out as a disadvantage compared to his colleagues in accounting and the like, and the restrictions imposed by his non-compete agreement only make it worse. If this engineer were to find it necessary to seek another engineering job, he might be able to do no better than a relatively junior position at a company in a distantly-related business. Even to get that job he'd have to successfully compete with younger applicants, some probably with more up-to-date and job-specific training.

As explained in The New York Times, "The growth of non-compete agreements is part of a broad shift in which companies assert ownership over work experience as well as work."[147] When an employer hires a senior engineer, it is doing so not so much to train him now in support of productivity later, but to take immediate advantage of what he knows. There is an arguable justification for a non-compete arising from the investment that the employer may make in training a relatively junior engineer, but for a senior engineer, less training is likely to be offered. It's far more likely that the dominant flow of information goes the other way, for example when the engineer uses his specialized knowledge to design a product. The employer has an unlimited opportunity to then take advantage of the engineer's knowledge as embodied in the product, but on termination, whether voluntary or not, the engineer who brought that knowledge into the corporation may be restricted from using it again until the non-compete expires.

One of the most pernicious aspects of many non-competes, at least to this non-lawyer, is their open-ended nature. The non-compete will typically refer to the employer's business at the time of termination. This can be particularly problematic when dealing with small companies, whose product lines and geographical scope of business may be expanding rapidly. It may even come to pass that the small company you joined is purchased by a vast conglomerate, and the non-compete that was once restricted to a narrow set of activities, is now being interpreted by the acquiring company to be much broader, encompassing virtually every professional job you are qualified to do. From the employer's perspective it would be a brilliant legal move to write a self-expanding non-compete; from the employee's perspective it can be disconcerting. This is something to discuss with your lawyer if you're concerned with the language you're being asked to agree to.

The employer may see a non-compete as having no downside, but there possibly is one. Given two job offers that are equal with the exception that one requires a non-compete, which one is the knowledgeable engineer likely to choose? Is the engineer likely to be attracted to states that don't aggressively enforce non-competes, and be deterred from taking a job in one that does? Established senior contributors probably risk the most by signing a non-compete; will they shy away from companies that insist on them? Conversely the absence non-competes in a specialty could make it more attractive, which to the employer's benefit could increase the number of trained prospective employees.

Employers can demand that you sign a non-compete not just when you begin employment, but at any time. Indeed, about a third of non-competes are signed after the job offer is accepted.[148] It can be easier for an employer to pressure you to sign a non-compete once you've resigned from your previous job and burned your bridges by turning down any other offers, so that in the short term unemployment appears to be your only other option. On receipt of a job offer, then, try to confirm before you formally accept it that it is complete, so that there won't be more obligations presented to you on your first day of work.

The demand to sign a non-compete could also come if you're working without one, and another company is interested in acquiring your employer; the acquiring company could ask for non-competes as a condition of making the purchase. If you're asked to sign a non-compete while you're already in a job, talk to your lawyer first. Find out whether it will be enforceable without specific compensation in return, and whether the employer could lawfully fire you for declining to sign.

An employer could require a non-compete at the time of an involuntary termination, as a condition for receiving severance pay. While severance payments after layoffs are common, they are not generally obligatory and the employer has wide latitude to attach conditions to them. Starr et al. found "that noncompetes are equally common in states that do not enforce noncompetes as they are in states that vigorously enforce them, raising questions about why employers use noncompetes and how even unenforceable noncompetes operate to change employee behavior."[149]

Can a non-compete ever confer advantages on the employee? Of course, if the employer won't hire someone without a non-compete, the job itself could be seen as the benefit. Starr et al. also found that employers who require non-competes at the commencement of employment may offer additional pay and benefits, although this conclusion applies across the workforce and not specifically to engineers.[150]

Exceptions to Non-Competes

With many lawyers extolling the benefits of non-compete agreements, and happily writing them on behalf of client employers, one might think that the lawyers would willingly embrace such agreements' restraints on their own behaviors. One would then be incorrect. It's considered unethical for a lawyer to either impose a non-compete requirement on another lawyer or even to agree to a non-compete on her own behalf. This is based on model ethics language developed by the American Bar Association, and so is generally the same from state to state. The following language, from Pennsylvania, is typical:[151]

<u>Rule 5.6 Restrictions on Right to Practice</u>

A lawyer shall not participate in offering or making:

(a) a partnership, shareholders, operating, employment or other similar type of agreement that restricts the rights of a lawyer to practice after termination of the relationship, except an agreement concerning benefits upon retirement or an agreement for the sale of a law practice consistent with Rule 1.17; or

(b) an agreement in which a restriction on the lawyer's right to practice is part of the settlement of a client controversy.

These rules of ethics may not be laws, but a court is likely to carefully consider them in determining whether to enforce a non-compete against an attorney. More likely, the attorney who requested such enforcement could be at risk of a sanction for an ethical violation.

In fact, there are any number of provisions written into state laws that exempt members of one profession or another from non-compete agreements. In Massachusetts doctors and certain other health-care personnel are exempt,[152] for example:

> Any contract or agreement which creates or establishes
> the terms of a partnership, employment, or any other
> form of professional relationship with a physician
> registered to practice medicine pursuant to section
> two, which includes any restriction of the right of such
> physician to practice medicine in any geographic area
> for any period of time after the termination of such
> partnership, employment or professional relationship
> shall be void and unenforceable with respect to said
> restriction; provided, however, that nothing herein
> shall render void or unenforceable the remaining
> provisions of any such contract or agreement.

It can be argued that lawyers' clients and doctors' patients have rights to choose their professionals and that such rights, as a public policy matter, should take precedence over the rights of their employers to restrict clients' and patients' choices. In other words, a case can be made that these rules and laws are intended not for the benefit of the practitioners, but for the benefit of those who seek their services. Under this logic, public policy need not restrict a corporation from imposing a non-compete upon an employee who does not directly serve a client or patient as a professional.

What, then, do we make of Broadcast Industry Free Market Act, Illinois Public Act 92-0496, effective January 1, 2002, which is reprinted in part below?[153]

> Section 10. Post-employment covenants not to compete
> are prohibited.

> (a) No broadcasting industry employer may require in
> an employment contract that an employee or
> prospective employee refrain from obtaining
> employment in a specific geographic area for a specific
> period of time after termination of employment with
> that broadcasting industry employer.

There's clearly more going on here than altruism for the public good. Why would it be in the public interest to ban non-competes for broadcast reporters and allow them for print reporters? The answer, of course, is politics. The affected employees intensively lobbied the Illinois legislature. They doubtless did so individually, but more importantly they did so collectively through the union that represents many of the employees covered, the American Federation of Television and Radio Artists.[154]

A World without Non-Competes

There already is one. It's possibly the most consistently innovative economy in the world, and it's called California. California law says "Except as provided in this chapter, every contract by which anyone is restrained from engaging in a lawful profession, trade, or business of any kind is to that extent void."[155] (An exception may apply to the seller of a business. As always, you should check with your attorney.) Employers elsewhere may argue that bans on non-competes are hostile to business, but the successes of the chipmakers and software companies in Silicon Valley, the movie studios of Hollywood, and the biotech companies in San Diego render the argument hollow. Tricky situations are possible here, such as for a resident of California who works remotely for an employer in a state that does enforce non-competes.

Non-competes are being challenged and changed in other states as well, and California is not the only state that significantly restricts them. And interestingly, in some cases the charge against non-competes is being led by investors – venture capitalists. These investors can find it difficult to staff critical positions at the startups they're funding if the people with the skills they seek are constrained by non-competes with their current employers, or perhaps if people with those critical skills are savvy and independent enough simply to decline to sign non-competes. Other states may allow non-competes but with statutory limits on how they can be enforced.

My research for this book turned up some executive-level non-competes, which generally show the effects of negotiation rather than coercion. For example, one such agreement specified by name the competitors for which the executive would be prohibited from working during its time in effect. This is the final indictment of the non-compete agreement and the manner in which it is applied: Corporations are quick to impose them on engineers, who may already be at a disadvantage with respect to portability of their skills and experience, but they are willing to negotiate terms with highly-paid senior managers, who presumably can do more damage by going to a competitor, because these senior managers are likely to have other credible job options

The value of a Non-Compete

Employers may expect their employees to sign non-compete agreements as a matter of course without receiving anything in return, but at the same time the business world recognizes non-competes as valuable assets that can be bought and sold. It will be interesting for us to look at the value of a non-compete from various perspectives, using some simple examples.

First we'll consider several examples of how an engineer might assign a value to a non-compete:

Example 1: An engineer is involuntarily terminated from his job but due to his non-compete cannot immediately accept other work in his field. He finds work as a retail sales clerk, at considerably lower pay. The value, in this case the cost, of the non-compete is the difference between the income (including benefits) he earns in retail and what he would have earned as an engineer, for the length of time the non-compete is in effect. This would also be the cost of the non-compete if the engineer is unhappy at his job and would like to take a similar job at a competing company but is prevented from doing so.

Example 2: An engineer is courted by a competitor with the offer of a more senior position and higher salary, but can't take the job because of the non-compete. The cost of the non-compete is at a minimum the difference between the two salaries, but is increased by the non-monetary value of the higher position.

Example 3: An engineer, not currently constrained by a non-compete, is pleased to find that his job search resulted in similar offers from two similar companies. The offer that includes a higher salary comes with a non-compete; the lower offer does not. How much greater does the first offer need to be before the engineer is willing to take it? In my experience many engineers are willing to agree to non-competes as a matter of course, so they are not likely to demand a large premium. In other words, the engineer in this case places a low value on the non-compete.

The calculations of Example 1 don't attempt to put a value on the engineer's inability to keep up with his field while working in another. We can see this as an opportunity cost that is difficult to quantify. In any of the first three scenarios the non-compete could cost the engineer thousands of real dollars plus the "irreducible" costs that can't readily be quantified.

From the employer's perspective, the calculations are different, of course.

Example 4: For the most part, companies place no monetary value on non-competes because they don't have to give up anything in return for them.

Example 5: A thought experiment may help estimate how an employer would value a non-compete if forced to do so: Suppose that there is a law in some jurisdiction that requires all non-compete agreements to specify a buyout amount; that is, either as a dollar amount or a percentage of annual salary that the employee or another party (presumably the employee's next employer) would have to pay to relieve the employee of his obligations under the provision. The specified or calculated dollar amount needed for buyout would be the value of the non-compete. The maximum could be limited by statute. This would give the employee at least a theoretical way out from under the non-compete's restrictions. Professional sports teams often use similar trading and buyout mechanisms for players under contract.

Example 6: To make the game more interesting, consider that through the course of years almost any company will spend some time as a net acquirer of talent and some time as a net loser. Suppose that government required all employers to conform to a universal non-compete buyout price, which when paid by an acquiring employer to the releasing employer would allow the employee to move to the former. In this case I think that employers would lobby to set the value low – less than in Example 5 because they know they'll be paying it as well as receiving it. I would expect the agreed-to amount to be well under the individual's annual salary.

Thus, my conclusion from these simple exercises is that many employers will continue to demand non-competes as a matter of course as long as they can acquire them at no cost. If they were ever somehow required to pay for a non-compete, the monetary value they would put on the asset would not be excessive.

For some other interesting economic models of the value of non-competes, again see Starr.[156]

Dealing with Non-Competes

Here is some business advice:

Have an exit strategy. Just as a venture capitalist always has an exit strategy when she invests in a company, you should have identified a number of business areas and prospective employers to which you could go without violating the non-compete. Since the scope of your employer's business may change over time, you should review this plan periodically. If "Plan B" requires some additional training, try to get it before you need it.

Avoid overspecialization. An old definition of an expert is someone who knows more and more about less and less, until he knows everything about nothing. If you get to the point where you know everything about some highly-specialized field, it can be hard to find a non-competing job.

Ignore the preceding paragraph. As you move through the ranks from junior engineer to senior, of course, until such time as your responsibilities shift toward project or supervisory management, you're going to have to become more proficient at your chosen specialty. It's notable that managers become more valuable to their employers as their experience broadens, but often engineers become more valuable as they become more specialized. At some point, leaving engineering for management or sales could become less a matter of career advancement than of self-defense.

Develop and Document Generic Skills. Even as you become a specialized senior engineer, you can still develop skills and credentials to serve you in an unrelated industry. If part of your job involves managing a project, try to get project management credentials from the Project Management Institute. If your job involves quality control, there's the American Society for Quality, particularly its Six Sigma certification program. You may be much better off presenting yourself to the marketplace as a project manager who most recently managed a project to design a new aircraft propeller, than as a propeller engineer who has some peripheral project management skills.

Ask your new employer to indemnify you. The upside of excessive specialization is that there may not be many people specializing in the same area you are. Even in a tight job market, you might get lucky and find an employer who needs exactly what you offer. In that case, the new employer might be willing to indemnify you, to help with legal costs, and perhaps even pay you for a time if you're prohibited from working. This is not a theoretical possibility – it happened to someone I know. The new employer might also be able to put you to work in an unrelated area until the non-compete runs out. By all means, though, don't try this on your own; have your lawyer assist you in developing an agreement with your new employer

Don't hide your employment restrictions: Your next employment prospect will likely want to know whether you're covered by a non-compete. Especially if you're applying through an on-line form, you may be forced to give a yes-or-no answer without the opportunity to clarify it, e.g. to specify its scope or the length of its remaining term. This is one more reason why non-competes should be avoided if possible. If you do have one, you should answer honestly. However, if you're under a non-compete that doesn't cover your prospective employer's business "no" could be an honest answer, depending on how the question is worded. Your strategy for responding to questions like this is something you may want to discuss with your lawyer, preferably before you're facing the on-line form

Negotiating the Non-Compete

If your prospective employer insists on having non-compete language in the employment contract, there may still be room for compromise. Every contract is at least potentially negotiable, including contracts presented to you bound in leather with the binder lettered in gold leaf, embossed with the corporate logo, printed on fine high-rag-content bond paper and bearing executive signatures written with gold-plated fountain pens.

The lawyer who prepared the non-compete agreement probably did so with the intent of drafting the strongest and most comprehensive such document that anyone could reasonably be expected to sign. Anything short of that, arguably, would have been legal malpractice. The wording of the agreement also reflects the power balance between the parties. In areas of life and commerce where the parties have roughly equal power, for example the purchase of a house, legal language tends to be more balanced. And certainly corporate managements would prefer not to have to negotiate a customized agreement with each new employee. But perhaps, with the advice of your attorney, you'll be able to successfully propose one or more modifications such as these:

Limitation in scope: The non-compete would apply only to areas you actually worked in; you would be free to compete in business areas that the employer is pursuing but that you did not participate in. For engineers working for smaller companies, this limitation would be particularly valuable when the non-compete contains "successors and assigns" language. Such language can act to transfer your obligations under a non-compete to the purchaser of your employer's business or its assets. Your concern is that this language not be claimed to automatically expand the scope of the non-compete to include any business engaged in by the purchasing company. Consider the possibility that you work for a small, highly-specialized employer that addresses one specific market niche. You might be comfortable with language that restricts your ability to compete with any activities your employer engages in, even if the employer achieves modest growth and enters a few new business areas. However, if your small employer is bought by a behemoth, the list of prohibited activities could be virtually endless. Your lawyer might tell you that a judge would likely refuse such a broad interpretation of the agreement, but how much money do you want to spend and how much risk do you want to take to prove her right?

By the way, even without successors and assigns language, there may be common-law principles that allow your employer's purchaser to acquire your obligation to comply with your current non-compete agreement, thereby greatly expanding its scope. This is all the more reason to work with your lawyer before you sign anything.

Enumeration of prohibited companies: Identify the employer's current competitors or customers; these would be the only new employers to which the non-compete would apply.

Term dependent on length of employment: The term of the non-compete could gradually increase, up to a reasonable maximum, with your tenure at the employer. The way that most non-competes are written, you could work for a company for one day and have a non-compete lasting for the full specified term, which could be a year or more.

Term dependent on type of termination: The agreement could apply only to your voluntary termination; that is, quitting, or termination for good cause such as a criminal conviction. Termination for other reasons would invalidate the non-compete or at least shorten its term. Be sensitive, though, to constructive termination – the wrongful imposition of intolerable working conditions – which effectively forces you to quit.

Termination on employer bankruptcy: The employee's obligations under the non-compete would terminate if the employer goes into bankruptcy. Your non-compete would then not become an asset to be sold off as part of the bankruptcy proceedings.

Termination on change of control: The employee's obligations under the non-compete would terminate if control of the employer changes, e.g., if the employer is acquired. Change-of-control provisions that provide compensation in the event a corporation is acquired are common in employment contracts made with executives. Such a provision allows executives to negotiate the sale of a company with reduced concern about their own financial situations. However, ordinary employees may also be at risk in the event of an acquisition.

Term dependent on severance pay: The agreement could specify a short term by default, which would be then be increased according to a stated formula if severance pay were offered and accepted.

Specification of a buyout amount: As discussed, this would be an agreement to a dollar amount or percentage of actual salary that would relieve you of the agreement's restrictions.

Your ability to actually get any of these modifications will depend on labor market conditions at the time, and as we know business tries to maintain a labor surplus. If you won't sign the non-compete, there are probably others who could do your job and would be willing to sign it. Many engineers won't be able to resist pressure to sign the document as presented, but if you're the exception you should try to get some reasonable changes.

If considering whether to fight your non-compete in court, be aware that information on court cases is available to the public. Just like the way an embarrassing picture on a social networking site can follow you forever, your public record as a "troublemaker" can live on in perpetuity. Conversely, your lawyer may be able to make use of records of cases similar to yours. An arbitration requirement, which commits both parties to a private process of resolution, could work in your favor.

And remember that, like viruses, non-compete agreements are constantly evolving, generally to your disadvantage. Language that is thrown out by a court in one month may be replaced the next month by a new version modified to be acceptable.

Living With your Non-Compete

If you have to accept the terms of a non-compete that you'd have preferred not to sign, don't despair. Courts in many jurisdictions have a history of being generally friendlier to employees than employers, so if some time in the future there must be a legal battle, you may have a chance of winning it. Possibly the best defense of all is to live in such a way as to make sure you can live in reasonable comfort for the term of the non-compete. While you're employed, take some courses to remain current, and keep watching for opportunities that wouldn't trigger the non-compete. Then be ready to come back roaring when the non-compete is up.

The time may come when you have no practical alternative but to seek or accept employment that might constitute a violation of your non-compete. You may be able to "break" your non-compete, but it's a difficult undertaking. Your employer (or ex-employer) is likely still in business and earning enough revenue to easily pay to outgun you in court. If your employer changed ownership, your non-compete is probably an asset under the new owner's control. You, on the other hand, may be out of a job and burning through your savings not only to meet daily expenses but potentially to mount a legal battle.

Bad things can happen to an employee who dares to break a non-compete. The New York Times tells the story of Keith Bollinger, a textile-factory manager, whose attorney advised him that his non-compete was probably unenforceable but was sued anyway by his former employer. Mr. Bollinger ended up "ruined" in his mid-50's by legal costs associated with his former employer's lawsuit.[157]

The Private Non-Compete

Anti-poaching clauses are written into agreements made between businesses and can affect you without your even knowing about them. Assume that your employer, Company A, gets a contract from Company B to supply something, and that in the course of delivering on the contract you meet and impress people at Company B. Company B decides it would like to have you as an employee. Anti-poaching language in the contract between the two companies may prevent Company B from employing you. The language is probably symmetrical, so Company B would equally be prevented from hiring Company A's employees. By their nature, anti-poaching agreements can be entered into at any time during your course of employment. They may be informal arrangements, not documented in contract language.

California is among a small number of states that all but prohibit non-competes, but that didn't stop several large employers in that state from allegedly creating their own anti-poaching pact that achieved an arguably similar result. On September 24, 2010 the United States Department of Justice (DOJ) filed a civil antitrust action[158] against California high-tech employers Adobe, Apple, Google, Intel, Intuit and Pixar. The allegation was that the named companies agreed not to cold-call each others' employees as part of their recruitment efforts. The DOJ complaint alleged that the defendants' actions "disrupted the normal price-setting mechanisms that apply in the labor setting" and "substantially diminished competition to the detriment of the affected employees who were likely deprived of competitively important information and access to better job opportunities."

The complaint alleged that this practice began with an agreement between Google and Apple made in 2006 or possibly earlier, that neither company would cold-call the other's employees. Over time the agreement was extended to other pairs of defendants.

The complaint was settled on March 17, 2011 with a final judgment that required the companies to refrain from "attempting to enter into, entering into, maintaining or enforcing any agreement with any other person to in any way refrain from, requesting that any person in any way refrain from, or pressuring any person in any way to refrain from soliciting, cold calling, recruiting, or otherwise competing for employees of the other person".[159]

After the DOJ complaint was filed several individual employees also filed suit in California state courts. These suits were consolidated, and granted class-action status on October 24, 2013.[160] On September 2, 2015, after some of the companies sued first settled individually, a federal judge approved a settlement of $415 million with Apple, Google and

Intel.[161] For detailed information on the case, see the settlement website.[162]

There was at least one similar lawsuit, also involving Pixar, claiming that a group of motion picture studios had created an anti-poaching agreement to hold down the wages of animators.[163] The suit was settled with a large payout.[164]

Non-compete agreements are governed by state laws; the federal government has little if anything to do with them. The complaint against the tech companies' anti-poaching arrangements was brought under the federal Sherman Antitrust Act,[165] which addresses anticompetitive practices in interstate commerce. The results alleged in the Department of Justice complaint are similar to the burden imposed on an employee who is required to accept a non-compete agreement as a condition of employment. Could non-compete agreements, when imposed by a company involved in interstate commerce, be successfully challenged under the Sherman Act? Such challenges have been made; I'll leave it to your lawyer to advise on whether such a challenge is likely to succeed in a particular case.

For more information on non-competes, see the website (what else?) non-competes.com, run for 10 years by a Chicago lawyer who specialized in them. That website is no longer being updated but as of this writing is still available. I have no experience with this site and of course, my reference to it does not imply endorsement.

Non-disclosure clauses

Few engineers would dispute the notion that an employer is entitled to maintain trade secrets – formulas, designs, marketing plans, and so on. At the same time, few would agree that engineers should be prevented from using the general knowledge of technology and business that they gained through working for one employer, in the course of their employment with another. Questions get complicated quickly, though. If the employee objects to an employer policy, or believes he may have been discriminated against or harassed in some way, is his knowledge of this policy or treatment the employer's trade secret? So where exactly are the lines, and is the proper location of these lines reflected in the language of the employment agreement? From the view-point of a society as a whole, do excessively restrictive non-disclosure agreements (NDA's) restrict technological progress and economic growth?

Where is the line when you are preparing your résumé? If you worked on the design of a product that was subsequently put on the market, noting that fact on your résumé may not upset your employer. But what if the product was never produced? It could be valuable for your employer's competitors to know that such a project was canceled, even if you don't tell them (and may not know) why. Engineers working on defense-related business may have particularly severe restrictions, although these would be imposed as much by law as by your NDA.

If you worked through your employer for a third party, does that third party not wish to be identified? If your employer won't disclose anything more than the dates of your employment, could the employer come after you for naming a customer or partner employee as a reference?

Sometimes employers aren't as clear as they should be about non-disclosure obligations that they have accepted on your behalf with suppliers or customers. Employees may then inadvertently violate an NDA that they have never been told about.

An overview of the issues surrounding NDA's in the Harvard Business Review asserts that they are "out of control".[166]

NDA's tend to be written broadly, in part because they have to be written to cover knowledge and information not yet generated, and in part because they reflect the power imbalance between employer and employee. It's helpful when the NDA explicitly excludes information you held when you started employment, including anything then publicly available, or which becomes publicly available during your work for the employer.

There are more questions than answers here, as there are many variations in the way NDA's are worded. I think an engineer's common sense can go a long way in determining what disclosures might breach a clearly-written NDA, but there may be times when you need to speak to a knowledgeable attorney. These times might include suspected violations of law, or perhaps ethical violations that are not explicitly illegal. A special case covers NDA's incorporated into arrangements which settle claims of sexual harassment or illegal discrimination. Some states have considered statutory exemptions to prevent NDA's from being enforceable against the person who made the claim, and some employers have voluntarily agreed not to seek to enforce them in such cases.

Non-solicitation (anti-poaching) clauses

This provision commits you to not soliciting the employer's clients, customers and/or remaining employees after leaving the company. These restrictions can prevent you from asking such contacts to join your new employer. A non-solicitation agreement can also be written to refer to soliciting business, and so overlap with a non-compete.

States may prohibit or limit the enforcement of non-solicitation clauses, and changes in laws and their application can come at any time.

Again, talk to your lawyer and make sure you understand how these and any other clauses can affect you.

Arbitration clauses

Arbitration is a process by which both parties to a dispute present their cases to an impartial arbitrator or panel of arbitrators, whose decision is then in most cases binding. It's one form of "alternative dispute resolution"; that is, alternative to litigation. There is also nonbinding arbitration, but it's less common. Arbitration can be considerably faster and less expensive than litigation. An arbitration clause is a provision in a contract that requires certain classes of disputes to be resolved through arbitration rather than litigation. While the parties in a dispute are free to mutually choose arbitration once the dispute has emerged, often the decision is made ahead of time through an arbitration clause imposed by the more powerful party. The arbitration clauses in employment contracts may also prevent a group of employees from cooperating in a class-action lawsuit against the employer.

In both consumer contracts and employee relations, mandatory or forced arbitration is increasingly common. The prospects of lower cost and greater speed are important, of course, but they're not the only reasons. The results of a court proceeding become public documents, and may provide guidance to parties that later find themselves in similar disputes. Conversely, the results of an arbitration decision may be private. This usually creates an advantage for the employer and the employer's advisers, who will have access to the records of previous arbitration decisions that they were involved in, but the employee will receive no such guidance. Further, subject to the particulars of the arbitration procedures being used, there is little right to discovery. For example, you might not be able to access the internal management memo that records the logic of the employer's decision to terminate you. There may be nominally a right of appeal, but the restrictions on using it are robust, and in practice arbitrators' decisions are almost always final. So if the neutral party isn't really neutral or if one party in the dispute acts unethically, there may not be much that the disadvantaged party can do.

Arbitration clauses in employment agreements often require the employer to pay the arbitration costs. Fine, but don't think you're getting something for nothing. While the arbitrator is pledged to be neutral, and I'm not suggesting otherwise, she is likely to see a lot more of your employer than she sees of you. An arbitrator who consistently rules against the party who pays her fees is not likely to see an increase in business from that party. So while I'm not suggesting that individual arbitrators will tweak their decisions to favor the party signing their paychecks, it is fair to note that the arbitration system as a whole comes with a built-in opportunity for bias. Further, the arbitrators' fees are not likely to be the only cost associated with arbitrating a dispute. As with a court case you may, for example, incur considerable expense in traveling to and from the arbitration venue if in-person sessions are required.

Like any other provision in a contract, arbitration clauses can be negotiated, at least in theory. Your lawyer might be able to help you get some of the more disadvantageous provisions of the clause changed, or add some favorable wording, such as a requirement that arbitration proceedings be conducted by teleconference. Also, remember that the arbitration clause only binds you, not any governmental agency that has the authority to act on your behalf. If you have reason to believe that your employer may have treated you in a way that is contrary to the law, talk to your lawyer before committing to arbitration; you may have other avenues through which to seek relief.

That mandatory arbitration is structurally disadvantageous to employees can be seen in the behavior of Google and other tech giants in 2018 with respect to claims of sexual harassment and assault.

Faced with employee dissatisfaction over the private resolution of sexual harassment claims, Google opted to drop the requirement for mandatory arbitration in favor of allowing the aggrieved employees to seek resolution in the courts.[167] Google was not the first to do so, but confirmed the trend. On March 3, 2022, President Biden signed the Ending Forced Arbitration of Sexual Assault and Sexual Harassment Act, which took effect immediately, giving employees the right to pursue their claims in court.

The extent to which the employer's right to arbitrate disputes is applied to other grievances such as age discrimination will depend on the company. It remains to be seen how many other companies, in Silicon Valley and elsewhere, will waive the arbitration rights that current employees were required to give over, and strike the requirement from their agreements with new employees.

In 2019 the House of Representatives passed the Forced Arbitration Injustice Repeal (FAIR) Act, not to be confused with the unrelated Federal Activities Inventory Reform Act of 1998. The 2019 FAIR Act would end forced arbitration in consumer and employment contracts. As of this writing, it has not been taken up by the Senate.

For more information on binding arbitration in employment, see the first part of this video from Cornell University's School of Industrial and Labor Relations:

https://www.cornell.edu/video/binding-arbitration-research-evidence-pros-cons (accessed March 30, 2021)

For arguments in favor of the arbitration process, see this position statement by the U.S. Chamber of Commerce:

https://www.uschamber.com/lawsuits/arbitration/us-chamber-letter-the-forced-arbitration-injustice-repeal-fair-act (accessed December 2, 2021)

How does the other half live?

Securities and Exchange Commission (SEC) regulations require that the employment agreements of publicly traded companies' chief executives be disclosed. By doing a little searching, you may be able to find executive employment agreements for your employer, a prospective employer, or other companies in the same industry.

To get a general idea of executive compensation, although not necessarily the details of the contract, look for the DEF-14A proxy statement. You can do a Web search using the term DEF-14A and the company name.

The website lawinsider.com collects examples of actual employment agreements, as well as other types of contracts.

Working with lawyers

I've referred in several places to the need to have a lawyer on your side. For employment-related matters this lawyer should specialize in, or at least have some interest and experience in, the recognized practice area of employment law. The reason that it's important for your lawyer to have such a focus is that laws change all the time, and even when they're not changing, their effects may change through court decisions. You may be able to get recommendations from people you know, or consult the website Workplace Fairness at https:// www.workplacefairness.org/find-attorney. I have not used this website and, as always, no endorsement should be inferred.

I'm not suggesting that you should ask your lawyer to represent you in negotiations with your employer, the way that a professional athlete or entertainer might be represented. This can be an expensive process for an engineer whose income is considerably less than his lawyer's. But it can still be helpful to sit with a lawyer for an hour or so and go over the agreement to make sure you understand it. Since she's a specialist, she should be able to advise you on how any questionable provisions in the agreement are being interpreted in your state's courts.

Before you do engage a lawyer, make sure you understand and accede to her billing policies. These start, but don't necessarily end, at her hourly rate. Some lawyers are more aggressive than others, and as an engineer on salary who may be expected to put in a lot of unpaid overtime on occasion, you may be surprised by how your lawyer bills. Anticipate, for example, being billed for quick answers to questions via phone, email or text. If your lawyer's engagement letter specifies that certain costs and fees are excluded, ask for a good-faith estimate of the total billed amount. Lawyers aren't necessarily accustomed to this, but you might even ask for documentation of expenses presented for reimbursement. After all, that's what your employer expects of you when you travel for business, and in some cases (do your research and/or talk to an accountant) documentation may make it easier for you to treat the costs as deductible business expenses.

The second most galling incident I ever had with a lawyer was when he couldn't immediately answer a question I asked him. He consulted with a colleague more familiar with the area of law under discussion, and then they both billed me separately for their time. As an engineering consultant, I don't believe I ever billed for acquiring or reacquiring specific tidbits of knowledge that my client could have reasonably expected me to already have. I certainly never made it possible for a fellow engineer not working on the project with me to bill my client for giving me a pointer or two when needed. The most galling experience, even though it involved only a modest amount of money, was when a lawyer received, on the same day, one-page responses to each of several filings that were related to each other but separate. Each such response was then forwarded to me in its own envelope, and I was billed for the time needed to stuff the envelopes (maybe also to make and file copies), as well as separately for each first-class stamp. I described this to a lawyer I knew socially who agreed that such behavior was not appropriate and felt, on behalf of his profession, embarrassed by it. So not every lawyer will act this way, and if you don't find out ahead of time you'll know soon enough whether yours is one who does.

If you question something in the employment agreement, you may be given an opportunity to speak with the employer's lawyer. Take the opportunity, but understand that the employer's lawyer has no obligation to be fair, whatever that means. The lawyer's obligation is to represent the employer's interests within ethical bounds. When she explains legal language to you, accept that her explanation is likely to lean toward the employer's interests, and also understand that it's unlikely that she has independent authority to change anything.

Some things that lawyers do

Here are some techniques that lawyers have used in contract documents or employed in face-to-face meetings with non-lawyers:

Dissociation: We know that sodium and chlorine are individually very nasty substances, but combine them in the right proportions and you get salt. This is the opposite: Related provisions are separated in space so that it becomes more difficult to understand them. Sometimes two contract provisions may seem to be not so terrible in isolation, but can combine to devastating effect.

Self-contradiction: Wording which appears favorable to you may be nullified by other wording elsewhere in the contract. For example, the beginning of a non-compete provision may have language like "In consideration of my employment by X, I agree that..." which may lead you to believe that if you sign the contract, X will be obligated to keep you employed. Later on, under an unrelated heading, may be language that says that the non-compete is not an employment contract and does not in fact obligate the employer to retain you.

The deceptive title: The title of a paragraph or section may not fully describe the latter's contents. This is similar to self-contradiction in that it requires careful reading. The most powerful provisions of a document can hide behind innocuous but irrelevant titles. When you see the word "Miscellaneous" as the title of a section of a document that you're being asked to sign, read it as "Things that can have a decidedly negative effect on my life".

Deliberate obfuscation: Of course, you should understand everything in a contract before you sign it. Have your lawyer explain anything that isn't clear, and then confirm her interpretation through Web resources. Watch out also for what appears to be common phrasing, but in fact means something different than you might expect. There may be key differences between the meaning of a term or phrase in common speech and its meaning in formal legal use. The difference could be as simple as a misplaced comma that changes the meaning of a sentence. There have been arguments up to the United States Supreme Court that turned on the placement of commas.

The Intimidator: Lawyers can draft agreements that they and their clients know have little chance of being fully enforced if challenged. We've seen this with non-competes. Many employment agreements, as a matter of fact, have "severability" provisions intended to allow the rest of the agreement to remain in force if a court prevents part of it from being enforced. It is good to know that a court may be able to do this. Also, I have read common codes of legal ethics, and have found no prohibition against the use of wording that the attorney knows is not likely to be upheld if challenged in court. After all, legislation or court decisions subsequent to the signing of the agreement could change the way certain phraseology is interpreted. Talk to your lawyer; don't just assume that the language of the agreement imposes immutable restrictions on you.

The throwaway concession: Non-competes often treat investing in a competitor as a form of competition, so will prohibit your doing so for the term of the agreement. There may be an exception for the ownership of a small number of shares in a publicly traded company. The exception may give you the impression that the drafter of the agreement was trying to be fair to both sides, but don't believe it. Consider that to take you to court for buying a few shares of a public company the employer would have to (a) know about it, (b) decide it's worthwhile to pay its lawyers to pursue the case, and (c) convince a court to actually enforce the provision. The likelihood that such a case would be taken to court, and that the employer would then prevail, is doubtless low, so the employer isn't really giving much away by offering this indulgence. Don't let gratuitous concessions like this convince you that the rest of the agreement is as reasonable.

A required statement that you understand the document: Agreements often end with language that says that both parties assert that they understand fully what the document means. Of course your employer understands it, having paid for it to be prepared. You, on the other hand, need to make sure that you don't sign it without in fact fully understanding it. It will always be difficult to say "but I assumed that..." when the meaning of a document is in dispute, but all the more difficult if you've agreed ahead of time that you did completely understand it. This statement may make the agreement more easily enforceable against you, and from the start it wasn't designed to be enforceable against the employer.

The strategic thumb: This was done to me at a deposition, in the course of an engagement as an expert witness. The opposing lawyer gave me a document to comment on, with his thumb obscuring some text that he didn't want me to read. I fell for it. Try to stay in control.

The final entry technically doesn't belong here since it's a management trick: Pretending that the employment agreement isn't such a big deal. It is.

A company may distribute a non-compete to its current employees written so that by signing it the employee agrees that "irreparable harm" would befall the employer if the employee were to go to work for a competitor within a year of leaving. The implicit threat would be to terminate an employee who refused to sign. Consider, though, that the employees would already be working without non-competes. If company management truly believed that serious harm would result from a former employee going to work for a competitor, then it would be an act of managerial malpractice to terminate such an employee for refusing to sign the non-compete. In fact it would arguably be malpractice to terminate the employee for any reason short of misconduct, since the employee would then be free to go to a competitor and thereby cause "irreparable harm".

Conclusion

Employers work hard to maintain a labor surplus in engineering fields. If you're not willing to sign their agreement, they'll likely find someone else who is. If you have to sign it, understand that employers have their lawyers, influence over state legislatures, and the attention of local politicians. Go ahead and sign it, and maintain your confidence and dignity.

Excessive specialization

As an engineer advances through the ranks, of necessity he becomes a specialist. The decision to do one thing, whether it is made by the engineer or his employer, is a decision to not do other things. Senior engineers in particular often find themselves disqualified from jobs that they could easily learn to do, because they can't immediately demonstrate knowledge, say, of a particular software package, type of product, or manufacturing process that they happen not to have worked with.

The general labor surplus in engineering is what makes it possible for employers to be this selective. They can define a rare combination of skills and experience to exactly meet their needs, but it will still be reasonably likely that such a combination can be found. Perhaps management already has a specific candidate in mind but has to go through the motions of posting the job. Human Resources staffers have, over the decades, become experts in defining open engineering jobs so narrowly that their preferred candidates are the only ones who qualify.

A lot of the initial screening of applicants' resumes is automated, and much of that screening consists of simple word matching. In part this is to reduce costs. It's also probable that the Human Resources folks don't know either the technical specifics of the jobs they're recruiting for or the meanings of many of the terms they get in response; little is then lost through the use of word-matching. But neither the HR staff nor the word-matching programs they use may realize when they are being presented with skills that are readily transferable, or faced with a candidate who is eminently trainable.

No judge likes to be overruled on appeal, and no HR staffer wants to be even partly responsible for a bad hire making his way into the organization. Strict adherence to a narrow definition of suitability will reduce the chances of such an occurrence, but it will also reduce the chances of bringing in a brilliant contributor with a fresh perspective. The good news for HR is that management will never know the opportunity cost created by this practice.

How does specialization affect people in other fields? Medicine is in a constant state of change from new scientific knowledge and new health threats, and law is always changing as a result of legislation and precedent-setting court decisions. A thought experiment: Consider the case of a physician who serves a patient population generally in need of statins – a well-known class of powerful cholesterol-reducing drugs. The costs of statins are decreasing as generics take over from patented compounds, but as of this writing annual US sales are still in the billions of dollars. With so much money at stake the question of which is the 'best" statin to take is of more than academic interest. (Not that academics aren't interested in money!) It may come to pass that an extended research study convinces this physician that many of her patients should be moved from one statin drug to a different one. So as her patients come in for routine exams, she conducts any necessary tests and switches prescriptions. No one will doubt this doctor's credibility or skills when she makes this decision, nor is she likely to lose patients to a doctor down the street simply because that doctor began prescribing the now "better" statin six months earlier. In other words, the addition of new knowledge to the physician's practice is incremental, not disruptive, and represents no material threat to her livelihood. The physician's career is not disrupted by the introduction of a new medicine.

Now consider surgery: Surgical techniques do change, and may require an investment in retraining and specialized new equipment, but no new appendages have been added to the human body since today's cohort of surgeons completed their training. A hand surgeon, for example, may well have to receive specific training on a new surgical method. But the training won't invalidate what she knows about how the hand works, and it will take advantage of the surgeon's existing knowledge store. Once again, change in surgery, as long as it's reasonably accommodated, is more incremental than disruptive, and to most practitioners is not likely to be a career threat.

What about lawyers? New laws have to be folded into the lawyer's practice, and precedent-setting court decisions may require her to change some of the advice she gives clients. Dealing with these changes is to one extent or another a part of every lawyer's practice. I would expect, for example, that intellectual property law is changing faster these days than personal-injury law. For both, though, experience is usually favored over currency of education. I have yet to hear of a client asking for a younger lawyer over a more-experienced one on the presumption that the former is more up-to-date on recent laws and court decisions. It may happen sometimes, but in engineering it happens every day as the norm rather than the exception.

Industrial economies have always benefited from cross-fertilization. Knowledge does not advance equally across all fronts and a lot of today's innovation can be traced to knowledge transfer between fields. One well-known example is the development of Velcro®, a brand of hook-and-loop fabric material. The product's inventor, Swiss electrical engineer George de Mestral, investigated how cockleburs were able to stick to his clothes and to his dog's fur, and used the results to help develop it. The "IBM Card", more properly the Hollerith card, was a mainstay of mainframe computing for decades. Its design was inspired by the punched cards used to control textile machines, beginning in the 1700's and perfected in the early 1800's at which time they supported fully automated operation of the machines. It seems likely to me that narrowly-focused hiring policies may be good for employers in the short term but hurt our innovation economy in the long term.

Vacation policies

Vacation benefits are not ordinarily addressed by employment agreements. Employers are free within broad limits (e.g., non-discrimination, state and local laws) to implement any vacation policy they wish for an employee not covered by a collective-bargaining agreement. Employee acquiescence is not needed. But it's worth noting the movement in business toward "permissive vacations" – replacing the traditional notion of vacation time accrual (e.g., 1.25 days per month worked) with the idea that you can take as much time off as you want as long as your management approves and you still can get your work done. Policies of course vary: Some companies require employees to take a minimum vacation time; some companies track vacation used and perhaps some don't; companies require varying levels of approval before vacation time is granted. Unofficial norms may also be different among companies, addressing questions like the total amount of vacation you're expected to take, how much notice to give and how to give it, and on what terms you're expected to be reachable.

Some issues will be the same as with traditional vacations, such as too many people wishing to take vacation at the same time. Engineers who support facilities that must run 24/7 are accustomed to further constraints. Sometimes, an entire factory shuts down for a facility-wide vacation, although this may be the time for engineers to come in and supervise maintenance and upgrades.

While permissive vacations are not ubiquitous, they are offered by companies ranging from small Silicon Valley startups to large ones like GE.[168]

Permissive vacations can be attractive, but they come with room for interpretation, as well as some risks for employees. If your workload is high, your employer may say that it's not a good time for a vacation, or you might self-regulate and not even ask. (In the traditional accrual system, especially if there is a limitation on carrying over vacation time from one year to the next, you may have a stronger claim.) If times aren't good at a company, the employee who says "I'm not too busy right now so I'd like to take some time off" may be volunteering to go from vacation to unemployment. If the company is in a chaotic state, for example if its sale to another organization is pending, there may be pressure for everyone to stay on the job. And outside of academia there aren't many "sabbaticals"; the engineer who decides to create one for himself is likely to attract negative attention, no matter what the company's official policy says. There are anecdotal reports of employees taking less time off under permissive systems, perhaps even competing with each other to see who can take the least and thereby show the greatest corporate loyalty. The vacation system, intended as a way to help reduce employee stress, may instead come to achieve the opposite.

Employees in a traditional, accrual-based vacation system often expect to get more time off as they accumulate time with their employers – perhaps one additional day per year worked for every year between 5 and 15. The permissive system removes that benefit of tenure and places senior employees on the same basis as their junior colleagues. It's not unreasonable to view the removal of a benefit of tenure as a form of age discrimination, as it eliminates an advantage previously granted to older employees. If employers can use the stick of non-competes to impel senior employees to stay, there is less need for the carrot of increased vacation time with seniority.

Permissive vacation systems work to the employer's benefit in another way that is not obvious and may not be explained to employees: Depending on local law, accrued vacation time may have to be treated as time worked, so that when an employee leaves the job, voluntarily or otherwise, the employer is obligated to pay for it. Without vacation accrual, there is no associated liability on the employer's books, and no need to compensate the employee if the vacation isn't taken.

To be fair, the permissive vacation system can be a welcome gift for someone who needs to take time off to care for a family member. Federal law (the Family and Medical Leave Act) allows for considerable time off, but it's unpaid.

At least one company in the United Kingdom rescinded its permissive-vacation policy when it found that employees weren't taking enough time off.[169]

Age Discrimination

Employees of age 40 to 70 are legally protected against age discrimination, but many engineers, among others, believe that they are subject to age discrimination nonetheless. Often, age discrimination results from "disparate impact" (as opposed to disparate treatment), wherein a policy that does not for an unlawful reason explicitly create a disadvantage for one group compared to another, nevertheless creates such a disadvantage in the way it is applied.

A good example of disparate impact can be taken from the first Title VII case (of the Civil Rights Act of 1964) on the topic to reach the U.S. Supreme Court:[170] When hiring laborers, the employer required applicants to have a high school diploma. The diploma requirement screened out more African-Americans than it did whites. Therefore, there was a disparate impact based on race, even though there may have been no discrimination explicit in the policy. Once the employees proved there was a disparate impact, the burden shifted to the employer to prove that the diploma requirement had "a manifest relationship to the employment in question." In other words, could the employer show that the knowledge and skills associated with a high school diploma were needed to perform the job of laborer? In this case, the employer was unable to do so and the Court decided unanimously in favor of the workers.

The Civil Rights Act of 1964 permitted claims of discrimination based on race, color, and national origin. It did not address age. Congress then addressed age discrimination in the Age Discrimination and Employment Act (ADEA) of 1967. ADEA prohibits age discrimination against those between 40 and 70. Generally, laws prohibiting discrimination based on age are weaker than those prohibiting discrimination based on race, color or national origin. An employer may assert that there is a "bona fide occupational qualification", such as for public safety, that justifies a discriminatory policy – for example requiring airline pilots to retire at age 65. In 2005 the US Supreme Court clarified that the disparate impact theory can be used in age discrimination cases; previously there had been differences between the ways lower-level courts applied ADEA. However, these age-discrimination cases are still difficult for employees to win. In a

2005 case a city gave police officers with less than five years service greater pay raises than more experienced officers. Obviously this had a negative impact on the 40-and-over group. The Supreme Court found that the city had based its decision on seniority (that is, officers with less seniority got larger raises), which was a "reasonable factor other than age". Therefore, the city won.[171]

A disparate impact is legally permissible if the employer can demonstrate a business reason that justifies it. At least the burden of proof is on the employer; it is not necessary for employees to prove the absence of such a justification.[172] This referenced case, decided by the Supreme Court in 2009, is particularly interesting to us as it involves engineers and other technical employees.

A recent development is the Protecting Older Workers Against Discrimination Act, POWADA,[173] which was passed on a bipartisan basis by the House of Representatives early in 2020. If passed by the Senate, this law would change existing doctrine that requires age to be the only reason for a determination that a practice is discriminatory, instead allowing the employee to prevail if age was merely one "motivating factor". Critics of the bill argued that it would have led to frivolous lawsuits and weakened protections afforded by other Federal anti-discrimination legislation. As of this writing the bill is under consideration in the Senate

Consider this possibility: An employer decides to trim its staff by simply dismissing the half of its employees who are over the median age for all employees. Obvious illegal age discrimination? Perhaps not, if the employer can demonstrate that the layoffs were based solely on salary, and can show that all of the older employees dismissed were paid more than all of the younger ones retained.

An employee may be able to use a claim for disparate impact in situations other than dismissal. Perhaps the employer offered a certain training course only to its younger employees, and then decided to terminate all employees who hadn't completed the training. A lawyer may tell you that a disparate treatment case is easier for the employee to win than a disparate impact case because the latter is more likely to require statistical analysis, expert witnesses and so on; but in practice the former will likely be difficult as well. The employer may assert a non-discriminatory reason for the action, and it could then be up to the employee, at considerable expense, to prove that the non-discriminatory reason was merely a pretext for discrimination and did not reflect a business necessity. Further, employees may be prohibited by their arbitration clauses from collectively pursuing their interests via a class-action lawsuit.

A savvy employer knows how to work around some of the limitations built into the law. A prospective employer probably won't ask for your age or date of birth, as doing so could arouse suspicions of intent to discriminate against older workers and could in fact be illegal. But is it age discrimination for the employer to ask the date of your degree? Generally not; that question may be acceptable since the date of the degree suggests how up-to-date your training is and there may be a legitimate business reason to favor candidates with more recent degrees, who in most cases will be younger.

In job recruiting, one might not see explicit references to age any more, but instead see terms with strong youth-related implications like "recent college graduate" or "digital native". Many older potential applicants are likely to just skip over such an ad without bothering to apply, so terminology like this is problematic.

A recent development in possible age discrimination is the use of social media by employers who wish to recruit from a specific demographic, a.k.a. younger people. The same techniques used to direct product advertising to a desired market segment can be used for employment ads as well. If an employer places a newspaper ad asking for people below the age of 40 to apply, most would see it as a clear case of legally-prohibited age discrimination. When the same employer instructs a social media platform to direct the ad only to people believed to be below age 40, the instruction won't be obvious to the platform's users and the legal situation may not be clear. In 2018, two legal scholars looked at several ways that employers could use social media to target recruitment ads away from minority groups and older workers, whether explicitly or implicitly.[174] In 2019, Facebook settled age-discrimination complaints alleging that its ad-targeting technology was used to illegally discriminate against older candidates.[175]

Employers are free to create disparate impacts in their employee benefits policies. In fact, it would be difficult to avoid doing so, since just about any benefit one could name will affect different groups of people in different ways. For example, dental insurance for retirees isn't generally covered by Medicare. Employer support for such coverage will be of value to primarily to older employees, while younger employees might be more interested in tuition reimbursement.

Employees may become suspicious, though, when the employer's expenditures for benefits visibly shift to the detriment of one group and the benefit of another. When your employer puts out a "benefits update", pay attention: Expect that improvements in benefits, typically of greater value to younger employees, will be announced with great fanfare, but that announcements of reduced benefits or increased costs, such as for medical insurance, will be disclosed properly but quietly.

Through interactions with a prospective employer you can get a sense of how it treats older workers. Look at the LinkedIn profiles of their employees – many include photos, as this is something LinkedIn encourages. Similarly, when visiting the employer's offices look at the faces: Is everyone young? (An appropriate age distribution for an established, large employer may be different than for a startup.) Ask about the corporate culture: Would you be expected to frequently socialize off-hours with your colleagues? That suggests an organization that is dominated by younger people. Try to speak with senior non-management employees to get their perspectives.

Of course, if you are a senior engineer you should not hesitate to share your specialized knowledge with younger peers, but be sensitive to undue pressure to do so. This could be a subtle signal that the employer is preparing to replace you.

Chapter 5: Managing your Finances

College costs

There's an abundance of material and assistance available on this subject, much of it at no cost. If you're preparing for college you should have no trouble finding it. I'm not going to try to duplicate it. I'll just say that with a societal emphasis away from corporate jobs and toward entrepreneurship, minimizing debt to maximize flexibility out of college will be of increasing importance.

Personal Finance for Engineers

What follows are my perspectives on some of the common financial decisions that people make through their working lifetimes, with particular emphasis on the needs of engineers. My words are offered as opinion only, derived from some experience and research but devoid of any claim or suggestion that I have relevant expertise or credentials. It would be best to consider my comments not as advice on how you should act, but rather as a discourse on how complicated a worker's financial life can get, and on the reasons why engineers will need some combination of their own due diligence and support from professionals to achieve a reasonable level of financial security. If you choose to engage a financial adviser, perhaps these remarks will suggest some questions you can ask to help select one.

First, I'll state three premises. I believe that these will apply to the majority of my readers. There's no shame in their not applying to you, but if they don't, then you will be able to get only limited value from my subsequent remarks.

Your principal goal is long-term growth at the cost of some volatility in the value of your investments. Typical goals of long-term investing include retirement income, home purchase, and college expenses for children. If your interest is more in short-term gains, cryptocurrencies, day-trading and the like, there's little to nothing here that is likely help you. Conversely, if any instability in the value of your investments makes you anxious, consider a bank savings account or a money-market fund. It may not be the most exciting or lucrative way to prepare for retirement, but you'll sleep better.

You are able to understand, and appropriately act upon, some of the basic mathematical concepts and techniques of personal financial management and investing. You are also able to read and understand financial documents, at the level of a well-educated non-specialist.

You are willing to devote time to understanding and managing your finances and to do some of your own thinking and make some of your own decisions, while not precluding professional advice where needed.

The need for financial planning

I believe that in significant ways, our country is becoming poorer. This statement may surprise those who have become accustomed to better smartphones every year, to reading about the contracts signed by celebrities in entertainment and sports, and to regular gains, even if not monotonic, in the stock market. It may also not be accurate according to the formal definitions of economics, but there is evidence behind it. We likely won't see a news headline exclaiming that "Americans became poorer last year" but that is the natural consequence of changes that are taking place continuously in our economy.

Pensions for private-sector retirees are becoming extinct, and even for public-sector employees they are often underfunded and under pressure. Employers will say with some justification that traditional defined-benefit pension plans (i.e., in which benefits are determined according to an established formula) have become untenable because people are living longer. Regardless of the reasons, the burden of financing one's retirement has shifted from the employer to the employee, along with the concomitant risks associated with investing money along the way. For an employee, providing for retirement income is one more demand to which time, effort and income must be allocated.

Social Security was never intended to serve as a form of pension or to replace a worker's salary. The Social Security Trust Fund is currently predicted to run out of money around 2037.[176] At that point the system will still have money coming in from active workers, but there will have to be some combination of tax increases and benefit decreases over time to keep the system in operation.

The cost of higher education has advanced faster than overall inflation. Student debt has impeded new graduates' purchase of new homes and the accumulation of wealth that it often engenders: In a 2017 survey, 86% of young millennials (born 1990-1998) and the same percent-age of older millennials (born 1980-1989), who were not then homeowners, reported that student debt was preventing them from saving for a down payment.[177]

Employers, certainly many employers of engineers, are relying on globalization to help maintain a labor surplus and thereby keep costs, i.e., U.S. employees' earnings, down.

There has been a sense in American society that the generation of workers coming of age now would be the first in memory to make less money than their parents. Research published in 2016 confirmed this: On the whole, Americans' chances of making more money than their parents have declined from decade to decade, reaching 50% for those born in 1980.[178]

Should it be a surprise that our nation is becoming poorer? We run trade deficits in the hundreds of billions of dollars each year. In other words, money – along with the wealth it represents – is leaving the country at a high rate. These are dollars that go to pay foreign workers rather than American workers, foreign taxes not US taxes, support foreign health care systems not ours, help build foreign infrastructure and not American infrastructure, and so on. The advantages of globalization are obvious; we see them in the form of low prices and broad product offerings at our favorite online and big-box retailers. The disadvantages are hidden. They may come in the form of higher social benefit costs, deferrals of public-works projects, increasing private and government debt and depressed salaries for US workers, including engineers.

All of this makes smart financial planning critical.

Some basics of personal financial management

> 1. Understand the pressures you will be subject to
> 2. Live below your means.
> 3. Take advantage of employer retirement programs
> 4. Evaluate job offers carefully
> 5. Beware of student debt
> 6. Be prepared for disruptions
> 7. Invest for the short, medium and long terms
> 8. Understand the investment landscape
> 9. Understand government programs for retirement

I'll look at each in turn:

Understanding Financial Pressures

Consider Social Security. Many Americans depend on it for at least part of their retirement income. The Social Security Act was enacted in 1935 and signed into law by President Franklin D. Roosevelt as a response to the widespread poverty brought on by the Great Depression. The original purpose of the program was to serve as a last safety net to shield retired workers from poverty, but over time it has become the default pension plan for many. Will the system be functioning as it is now when today's young engineers retire? Many believe not: A 2014 Pew Research poll found that about half of millennials and those in Generation X don't expect to receive any Social Security benefits at all. Only six percent of millennials expect to receive benefits at current levels.[179]

You have a Social Security (Account) Number and may periodically receive personalized statements, but your Social Security account is not like a bank account; it does not represent invested money to which you or your heirs have a legal claim. Rather, your payments from Social Security depend on the year of your birth, your work history and the age at which you (and possibly your spouse) begin to receive benefits. There is no guarantee that you will receive more than or even the same amount of money that you put in. In fact, in some cases (e.g. you've already reached the maximum contribution amount for 35 calendar years) you are guaranteed not to receive any additional benefit in return for the payments you put into the system as you continue working.

A study published by the Urban Institute in 2018[180] projects the gap between payments and benefits for a variety of situations if policies remain as they are. If the authors are correct, many retirees will receive less from the retirement system (Social Security and Medicare) than they will have paid in taxes.

Further, the retirement system and the tax system as a whole constitute a "complex system" in the technical sense that it has many parts that interact with each other in ways that may not be easy to discern. The practical result is that you are likely to receive less from the system than a casual reading of the rules would suggest. For example, Social Security payments up to a maximum percentage (85% at this writing) are subject to taxation, as a function of your income. Your calculated income includes your Social Security benefits themselves as well as tax-exempt interest income. This was one of the changes enacted in 1983 to "save" the system by reducing its real payouts. Almost literally, the federal government gives you money due you with one hand, and then takes back a part of it with the other hand.

Even the simple question of when to start claiming benefits can become complicated, especially for a couple. Given that an individual can claim Social Security on his or her own account at any time from age 62 to age 70 (8 years x 12 months = 96 options) a couple would have almost 10,000 choices with respect to their own accounts, and many more from the ability of one spouse to make a claim on the account of the other. It's a common strategy to delay the start of benefits to increase monthly payouts, but some critics have suggested that the government may someday decrease the value of such a delay as a way to "save" the system once again. You will need to spend some time to find the optimal claiming strategy as a function of your own circumstances. Some but by no means all financial advisers have expertise on this subject. Some advisers specialize exclusively in Social Security and fortunately there are a lot of resources on the Web, many available at no cost.

There's a relationship between salary compression and Social Security earnings that does not work to the benefit of most engineers. Once an individual's income reaches the Social Security maximum in a year, Social Security taxes stop but Medicare taxes continue. The Social Security maximum increases every year. When your Social Security payment is calculated (that is, Social Security's payment to you), your income in each year is compared to the same year's maximum income subject to taxation. So if the Social Security maximum increases faster than your earned income, from year to   you will fall behind.

Remember that your 35 "best" years are used to calculate your Social Security payment amount. Your first 35 years of full-time work may be your best 35 years, in which case by continuing to work you will not raise your "primary insurance amount". (You can, though, increase your annual payment from Social Security by delaying the start of payments, at present up to age 70.)

To get a sense of how much Social Security is likely to cost you and do for you in the future, you might want to look at the history of three important numbers: The trend of maximum income subject to FICA tax (Federal Insurance Contribution Act) for Social Security, the cost-of- living adjustments (COLAs) granted to current recipients of Social Security, and general year-to-year inflation. In the table in Appendix A, the FICA Maximum column doesn't reflect Medicare contributions, for which there is no maximum income, and the COLA column does not reflect changes in Medicare costs which can take away some portion of the benefit increase. The CPI column is the Consumer Price Index for Urban Wage Earners and Clerical Workers (CPI-W) for January of the year specified, as published by the Social Security website, and the Inflation column is the calculated year-over-year increase in the CPI.

While cost-of-living increases roughly track inflation, the maximum FICA wage has increased much more quickly. A given value of annual earnings represents a smaller fraction of the FICA maximum as the latter goes up. As time goes on fewer and fewer workers are likely to reach the FICA maxima for 35 years to qualify for the maximum benefit. Social Security benefits received by retired workers are therefore likely to decrease relative to the maximum benefit for their year of retirement.

Since Social Security was designed to be a pay-as-you-go system, the ratio of contributing workers to retirees gives a good index as to its health. That ratio steadily declined over time to 3.2 workers per retiree in 1975 with the latest available ratio for 2013 at 2.8, the lowest value on record.[181] It's likely – all but inevitable – that future changes to "fix" the system will include some combination of higher FICA taxes, lower (including delayed) payouts, and additional income tax. Finally, you should know whether your Social Security payments will be subject to state taxation. State governments are under financial pressure and are always looking for ways to increase revenue. Several currently tax Social Security benefits.

Live below your means and invest the difference

Unless you participate in a pension plan, your employer is not putting away money for your retirement; the most it will do is match your own retirement contributions up to a limit, and for an employer even that is optional (unless required, e.g., by collective bargaining agreement). The federal government isn't putting away money for you either, really; in the pay-as-you-go Social Security system the government is sending your Social Security taxes to current retirees and in return promises you that it will require future workers to send some of their earnings to you. Some commentators have compared Social Security to a Ponzi scheme[182] – illegal for private enterprise, but acceptable as a government program. The simple fact is that the entity that is responsible for your present and future financial health is YOU.

Fortunately, starting salaries in engineering are high compared to other fields requiring four-year degrees. For most engineers, living in most areas, it should be possible to establish a comfortable lifestyle while generating surplus income that can be made available for investment. Making investment a priority may mean living initially with roommates, buying a used car rather than a new one (or not buying one at all if you can manage it), and going easy on vacations. Learn how to cook or at least to microwave, and favor clothes that you can maintain yourself, without dry cleaning. Invest what you save in a properly diversified portfolio, making sure to limit your own employer's stock to a reasonable percentage. This is to help control common-mode risks, wherein a problem at the company you work for can both cause its stock price to go down and put your job in jeopardy. Your voluntarily modest lifestyle will be worth it, not only once you retire, but along the way. Money in the bank can help you meet financial challenges, such as a job loss, with some equanimity. It can also help keep you from being put in a position where you have to accept a bad job deal when it's the only deal available at the moment.

You might also invest some of your surplus in paying off your student loans early. Being out of debt is a positive thing, particularly so if you're interested in creating a startup, or if you want to work for a startup which emphasizes payment in stock options or restricted stock over cash. Being out of debt may also help improve your credit rating, which could then result in lower interest payments for a car loan or mortgage.

Take advantage of employer retirement programs

Everybody should know about 401(k)'s and IRA's. Contributions to traditional (non-Roth) 401(k)'s which you make through your employer, and traditional IRA's (Individual Retirement Arrangements or Individual Retirement Accounts; both terms are used) which you make on your own, are not taxed when they are made. In other words the amount of these contributions, subject to calendar-year limits, is subtracted from your income for the purpose of determining your income tax. Money put into these accounts then grows tax-free until the time comes to withdraw it. The 403(b) plans that employees of tax-exempt organizations can get are similar. These traditional plans should be strictly used as retirement assets, since early withdrawals if even allowed are subject to stiff penalties. These plans, in contrast to defined-benefit pension plans, are often termed defined-contribution plans since you know what goes in but their returns can't be known ahead of time.

Some plans do allow the holder to borrow against the account value, and there may be specialized penalty-free early withdrawal options available under some circumstances. Do your research or check with a knowledgeable financial planner.

401(k) accounts (the reference is to the section of the Internal Revenue Code which authorizes them) have largely replaced pensions, particularly in the private sector. By transitioning from a conventional pension to a 401(k), the employer eliminates its commitment to fund the employee's retirement, over time eliminates the necessity of maintaining pension funds (which can make it easier for the company to be taken over), and eliminates the risk that under-funding the plan or making poor investment decisions can lead to financial instability. An employer is not obligated to offer a 401(k) or to match the funds you contribute to it except perhaps by contract with a union, but today 401(k)'s are commonly offered and employers often match employee contributions up to a specified maximum. Employee contributions are always immediately vested, and the associated investment earnings belong to the employee as well. However the employer match may not be vested immediately, leaving some portion of the account subject to forfeiture if the employee leaves, whether voluntarily or not.

Under a 401(k) or similar account, the employee has the both the opportunity and the burden of choosing how his retirement money is invested. It is a common criticism of these accounts that employees don't always have enough financial education to be able to do so successfully, and a further criticism is that 401(k) plans can confuse employees by offering too many, rather than too few, choices. (Be aware that, just as ballot position bestows an advantage to the top-listed candidate in an election, the funds named at the top of a plan's list may not be the best for you.) If this is true it may be getting worse, as some plans are expanding to include annuities and private-equity investments. In my experience employers offer a lot of "information" but little "advice"; with the former the employer takes on little liability for bad investment decisions. To get the best value from your employer's 401(k), you have the responsibility of getting educated on the characteristics of all of the investments offered, or to arrange for professional advice independent of your employer.

You should also know that employer-sponsored retirement accounts can be expensive investments. Just as any other investor in a mutual fund, you may help pay for fund management. You may also pay fees to the company or companies that select the mutual funds that you're offered, arrange for contributions to flow to the mutual fund managers, and generate consolidated reports. Generally, smaller plans – plans with a small number of participants and corresponding small amounts of money under management – are the most expensive. An employer-oriented website found that the smallest plans have expenses averaging more than 3.5 percent.[183] Fund fees combined with the other fees that are often applied to these accounts (do a Web search for "401(k) fees") can take a surprisingly large bite out of your retirement savings. To the extent that there's an employer match you should still be able to do better in an employer-sponsored account than you could on your own, but it's not guaranteed. You should be aware of the costs applied to your retirement plan and the extent to which your employer covers them.

Those who exercise control over 401(k) plans are considered to be fiduciaries, which means that they are required to put the employees' interests first – for example, by selecting a lower-cost mutual fund over a higher-fee equivalent. Until 2012, service providers could hide their costs, but now disclosure is required.

Conventional mutual funds have been dominant in 401(k) plans but exchange-traded funds (ETFs) should be considered when available. Their expenses are generally lower than those of conventional funds, and over time the difference can be significant. You might want to talk to your employer if there are ETFs that mimic the performance of some of your plan's mutual funds but at lower cost.

When these tax-deferred accounts were introduced, they were often promoted as vehicles for reducing the total tax burden on the money invested and its subsequent gains. This was to be the result of both tax-free accumulation and the anticipated lower tax rate on withdrawal. It was commonly presumed that the account holder would be retired and earning less, and thereby be subject to a lower marginal tax rate (i.e., be in a lower tax bracket), when the funds were withdrawn. Certainly assets in these accounts grow without being immediately taxed, but the effective rate on withdrawal may not be lower. An engineer may be contributing money to 401(k) accounts over the course of a working career of 45 years or more; who is to say what the tax rates will be at the beginning of retirement, or how they might change during retirement?Tax rates and the fundamental structure of the taxation system (e.g., what's taxable and what's not) are set by Congress and state equivalents through a political process that reflects, among other things, government spending. At this writing, government spending, at least at the federal level, is increasing dramatically.

In considering the part that these accounts should play in your retirement planning, you should also be aware of two important rules that govern how some accounts are taxed. First are the "Required Minimum Distributions," RMDs, that force you to withdraw money once you reach a specified age. Withdrawals under RMDs (and from your traditional retirement plan generally) are taxable as ordinary income. If there's a lower tax rate available for capital gains, your retirement-plan earnings won't qualify for it. As of this writing the rates on capital gains are lower, but there have been proposals to eliminate that advantage at least for some taxpayers. In addition (see how the different programs and rules interact!) when RMDs increase your taxable income you may also be put into a higher tax bracket, have more of your income from Social Security become taxable, and possibly see an increase in your Medicare insurance costs since premiums under Part B are a function of income.

Beware of Student Debt

Outstanding student loan debt in the United States exceeded $1.5 trillion at the end of March 2019.[184] You know that when a college financial aid office says "we will meet your financial needs" they may mean that they expect you to borrow a large amount of money. It's not unheard of for a borrower to spend 25 or even 30 years paying off his undergraduate debt.

Money, in the form of freedom from excessive debt, may not buy you happiness but it will buy you options – the option to more comfortably work for a startup, accepting a low salary for a while in return for the possibility of a greater return later, the option to work for a nonprofit at a low salary, or the option to go immediately to graduate school without creating an overwhelming debt.

But there is some good news here for engineers: First, with the field's relatively high starting salaries we may be in a position to significantly accelerate repayment. Second, solid engineering programs, some of them world-class, can be found at state universities, on the whole less expensive than private institutions. State universities have generally been increasing their reliance on tuition and fees as public support has decreased, but many states still see the strong connection between engineering education, job creation and increasing tax revenues.

One more point about debt: Not all are the same. Debt incurred in the purchase of an asset that then appreciates, for example a home in most cases, helps to increase the return on the money invested and paid for debt service. The opposite occurs with depreciating assets, such as, traditionally, automobiles. Salary compression can similarly constrain the return on money borrowed for an expensive education when less expensive alternatives are available.

Be Resilient

Employees generally are advised to have emergency funds, enough to support themselves and their families for some length of time in case of a job loss. This advice applies in particular to engineers, who in the normal course of events may see their projects finished, their products go into production (in China, perhaps), their employers' business disrupted by technological or other change, or their industries decline. A typical goal would be to be have the funds to cover six months of expenses.

Involuntary job loss typically qualifies you for unemployment benefits and you may be able to draw severance pay, although the latter can come with conditions. Federal law (the Consolidated Omnibus Budget Reconciliation Act of 1985, COBRA) gives most employees the right to maintain their membership in their employer group health insurance plans, including dental plans, for a limited time after job loss, but there may no longer be an employer subsidy. Your Human Resources department should be able to give you the details.

Invest for the long, medium and short terms

Suitable short-term investments are liquid (can be easily converted to cash) and don't fluctuate too much in value. Investments having this characteristic can be converted into cash when needed without much loss of value such as could occur if you have to sell a stock when its price is low. These investments are what you need to get through short-term disruptions as described above. Short-term investing is also suitable for saving for a down payment for a house, for example, unless you're willing to wait for a reversal in a declining stock market before you buy.

Investing for the longer term is what you need for retirement. While the long term is comprised of a succession of short terms, long-term investments can withstand some short-term value fluctuation as long as you'll never be forced to sell a large chunk of your investment at a low price.

Medium-term investments are then intermediate between the two extremes, likely showing significant volatility, but over an intermediate length of time e.g., a few years, can be expected to show good returns. These may, for example, be best for an individual already in retirement. Investment portfolios may be required to support a combination of short-, medium- and long-term goals, and may therefore use a combination of all three time frames. Investments may be held in different "buckets" each managed for its corresponding purpose. Buckets may also be designated for specific planned expenditures such as the down payment on a home or college expenses.

Understand the Investment Landscape

There is a small universe of asset classes into which funds can be invested. Most investment plans emphasize securities, which may be either stocks (an equity holding in a company) or bonds (the debt of a company, or of a government or governmental agency). You may also be able to invest in real estate, directly or through a real estate investment trust (REIT). Additional asset classes may include commodities, collectibles, and currency including cryptocurrency, as well as a few others. From a small set of these classes, the investment industry has developed an astonishing number of things it can sell you – mutual funds that combine many individual securities, exchange-traded funds (ETF's) that are effectively a specialized form of mutual fund, annuities, bonds that can be converted to stock, options to buy or sell a stock at a specified price, preferred stock, collateralized debt obligations that aggregate income-generating securities such as mortgages, and so on. And new products are frequently developed. With the exception of mutual funds, whose principles are usually simple to understand, the further one gets from the underlying securities the more obscure the investments can become, and the more difficult it can be for non-specialists to understand them, understand their risks, decide how much money to allocate to them, and choose the best ones.

A diversified portfolio allocates investments not only into buckets corresponding to various time horizons, but within each bucket limits exposure to individual companies, areas, or industries. In other words, a large allocation to one oil company would not be diversified, splitting the allocation among several would be more diversified, and even more diversified would be an allocation that invests in renewables and other non-traditional forms of energy as well as fossil fuels. If investing in mutual funds, be aware of the possibility of overlap in their holdings; your portfolio may be less diversified than it appears.

In addition to choosing how to allocate your investments, you have some choice in how they will be taxed. Indeed, certain investments in government debt can be exempt from federal taxes, state taxes, or both.

Taxable and Tax-Advantaged Investments

In the early 1970's, Congress passed legislation to correct what some saw as abuses in the pension system, such as when a company's rules required 20 years of service to qualify for a pension (vesting) and then it terminated employees just short of that mark leaving them without pensions. The Employee Retirement Income Security Act of 1974 (ERISA) doesn't require employers to offer pensions, but does set standards for the pensions that a company could choose to offer. Today, few private-sector employers offer traditional pensions to new employees.

Section 401(k) of the Internal Revenue Act was revised in 1978 to allow employees to receive deferred compensation tax-free. This change enabled the first 401(k) plan to be offered, by The Johnson Companies[185] in 1980. Briefly, the traditional 401(k), now a staple in retirement planning, allows the employee to make before-tax contributions (that is, the contributions are not taxed in the year in which they are made), supplemented by an employer match to the extent the employer chooses, both of which then accumulate tax-free until withdrawal.

Similar to this is the Individual Retirement Account (or Arrangement), or IRA. An IRA has many of the characteristics of a 401(k) but is based on an arrangement between an individual and a sponsor, typically an investment company, and does not involve the individual's employer. Often, employees who leave a job before retirement age will transfer ("roll over") their 401(k) holdings into an IRA; if done in strict accordance with the rules this is not a taxable transaction.

All money in the traditional 401(k) or IRA — accumulated employee investments, employer matches and investment earnings — is taxed upon withdrawal as ordinary income. That is, the withdrawn amount is not treated as a capital gain no matter how long it's been invested. Tax laws normally give you a break on money that you're earned through capital gains; this encourages investment over spending and in part compensates you for putting your money at risk through an investment with an uncertain return. In fact, many countries don't tax capital gains at all. But withdrawals from tax-deferred accounts are always, under Federal law as of this writing, taxed at the full ordinary income rate even if much of the money earned within the account is originally from capital gain. Quarterly and online reports from your 401(k) or IRA custodian don't include the effects of taxes, thereby tempting you to over-estimate how much you will actually be able to receive from the account. You may wish to emphasize growth-oriented investments in your taxable accounts to take advantage of favorable capital-gains treatment, and dividend- and interest-paying investments in tax-deferred accounts particularly when interest rates are high. This gives you the advantage of diversification while maintaining some of the benefits of tax-advantaged capital-gain treatment.

Tax law changes introduced in the late 1990's give workers options that get around some of the unfavorable aspects of the original tax-deferred accounts. These are, of course, the Roth plans. Under a Roth plan – either a 401(k) or an IRA – contributions are taxed along with the rest of your income in the year that they are made, but then can grow tax-free until distribution (withdrawal). On distribution, Roth proceeds are not directly taxed and so won't potentially put you in a higher tax bracket. They also won't increase the tax on your Social Security benefit, which can happen with a traditional tax-deferred plan.

Be aware of the significant differences between the Roth IRA and Roth 401(k), at least as of this writing. For example, Roth 401(k) accounts are subject to required minimum distributions, but Roth IRA's are not. Similar to the case for traditional retirement accounts, to avoid a penalty for early withdrawal (absent certain hardships and other pos-sible exceptions) you must be at least 59 ½ years old and have held the funds in the account for at least 5 years. However you can withdraw your contributions from a Roth IRA (contributions only, not gains) at any time without tax or penalty.

The Roth options may be particularly attractive to young engineers who expect their earnings to increase and/or tax rates to increase as their careers develop. Paying some tax now may be better than paying considerably more tax later. So should everyone use a Roth account over its traditional equivalent? Not necessarily. There is an argument to be made for dividing your retirement savings between traditional and Roth accounts, especially as you approach retirement. Remember that you have to hold on to money in a Roth account for a while before you can withdraw it tax-free. Further, both types of accounts could become tempting targets for some future Congress seeking to increase tax revenue as the national debt continues its dramatic increase. Since the proceeds of traditional accounts are already taxed at your highest ordinary-income rate, a change to overall rates might be required to increase the taxes on them. Conversely, Roth proceeds could be exposed to taxation through a relatively small tweak to the tax code that would affect no one except Roth account holders. Either or both could happen, but a change focused on Roth accounts might be politically more palatable, especially if only a small percentage of the population has them.

Since we can't predict the future tax landscape, one could do worse than to put some money into investments that have at present no explicit tax advantages. Ordinary bank and credit-union savings accounts (physical or online) aren't the most exciting places to put money, but they may be the last to be hit with new taxes. And they're typically protected by federal insurance, either FDIC (Federal Deposit Insurance Corporation) for banks or National Credit Union Share Insurance for credit unions.

By investing a set amount of money into your retirement account according to a set schedule, e.g. monthly, you engage in what is called dollar-cost averaging. When you dollar-cost average into a volatile investment, for example an equity mutual fund, you purchase more shares when the investment's value is low and vice versa. Over time this strategy will push down the average purchase cost of a share of the fund.

Be careful about putting too much money into your own employer's stock. A company, for example, may match your contribution with its own shares. If the stock zooms upward, of course, this is wonderful. But if the company does poorly, you can face the risk of losing a big part of your retirement savings at the same time you lose your job. Learn the rules and make sure your investment in your employer's stock doesn't become too large a part of your portfolio.

One of the reasons why people often move their savings into an IRA when leaving a job, even when they would have been able to leave the funds in their employer's 401(k) plan, is that an IRA typically offers many more choices of investment. These choices may include funds that are near equivalents to the 401(k) offerings but less expensive. Over time, the difference can matter.

Once you've decided how to allocate money going into your plan, you will have a second set of decisions to make, concerning what you do with the money already there. One option is simply to do nothing; leave the money in each individual fund to accumulate as it will. But it can be better to balance. Balancing means that you decide on an ideal mixture ("balance") of asset classes, for example 50% stocks and 50% bonds in dollar value, and then either periodically or on an event basis (typically when the portfolio becomes sufficiently unbalanced) move funds among the investments to restore the balance desired. This can be effective because of rotation in leadership positions: One year domestic large-caps (capitalization over $10 billion, the biggest companies) may return the best performance, the next year it may be international bonds, and so on. By balancing a portfolio you take some money away from previous winners and direct it to investment classes that may flourish in the near future. You can balance your account manually, set up automatic periodic balancing in some cases, or select funds that do this routinely, within the funds themselves. These latter are "balanced funds". Balanced funds favor mathematics over judgment, which may be a good thing. However in principal a balanced fund will go to zero if any one of its constituent funds goes to zero. This isn't something we expect, of course, but if you're balancing your accounts yourself you can decide when too much of a good thing can become a bad thing. Further, self-balanced portfolios can have an advantage when it comes time to withdraw funds. You can withdraw money preferentially from stocks when you think the stock market is high and from bonds when you think it's low. With a balanced fund you can't control this and in fact could be withdrawing more from stocks when the market is low, since the fund may have been buying stocks to maintain its desired balance.

If your time frame is long enough you may wish to direct a relatively high portion of your investment plan contributions into the most volatile assets within your portfolio. Then you can decide to rebalance the portfolio when the volatile investments have done well.

"Target-date funds" are balanced funds wherein the allocation among investments becomes less aggressive over time. This strategy favors more volatile investments initially, moving over time to less-volatile investments, typically bonds, as the target date nears. The fund's published target date specifies the date at which the fund managers assume its account holders will retire. The sponsors of these funds typically offer funds with a variety of target dates to accommodate workers of different ages and retirement objectives.

You can mimic the operation of a target-date fund by changing your desired balance over time, away from more volatile investments toward more stable ones. If you're interested in doing any kind of balancing with funds already in your retirement account, check your retirement plan rules for any restrictions on trading frequency. Alternatively, instead of moving money among funds you can direct more of your periodic contributions to investments whose values are below the percentages you're trying to maintain. There should be no restriction on how your contributions are directed to different funds, although some plans may limit the frequency of your allocation changes.

Index Funds

Index mutual funds and similar exchange-traded funds attempt to track a specified market index. The first index fund was registered in 1972. Such a fund is passively, not actively managed; fund managers don't frequently trade stocks or bonds in pursuit of the greatest possible return. Thus an index fund is not very expensive to run, and in a competitive marketplace these savings are passed on to the fund's investors. In fact, there are funds that charge no management fees, although of course the fund managers are compensated in other ways. Even without active management it's not uncommon for index funds to achieve better results than their actively-managed competitors. The prospect of below-average cost combined with average-or-better gross returns makes index-fund investments attractive.

You can expect funds that track the same index to show similar results, but they may not be identical. For example, one index fund may try to own all of the stocks in an index, while another might own the stock of one company to mimic the behaviors of similar companies which the fund doesn't own. The latter may have a higher "tracking error". And make sure you understand the index that a fund is tracking; the better-known indices are very broad but there are others that are narrow, e.g., they may track a narrow industry segment, similar to "sector funds" which focus on banks, utilities and so on. (Sector funds themselves may be managed actively or passively.) Perhaps these specialized funds should play at most a minor part in your portfolio.

A portfolio with index funds tracking several different indexes can still be balanced, either automatically or manually. Watch for overlap; try to avoid investing in indexes that invest in part in the same companies. Index funds have been criticized for being too successful – they've grown so large that a small group of fund managers can exercise voting control over huge companies, and they can distort the usual relationship between corporate performance and stock price. Large active investors have for their part been criticized for the opposite – jumping in and out of holdings based only on short-term criteria.

Annuities

Funds reserved for future use can be used to purchase annuities instead of being invested in the conventional way. An annuity is a contract with a financial institution, typically an insurance company. The issuer of the annuity promises, in return for one or more payments to the annuity in the "accumulation phase", a stream of payments from the annuity later in the "annuitization phase". The largest financial institution offering annuities is the United States Government through Social Security and Medicare, so a majority of the US working population participates in at least one. Workers in defined-benefit pension plans, becoming rare in the private sector, are also participants in annuities.

We understand insurance as a way to reduce the risk associated with an undesirable event beyond our control, by sharing that risk with others. An insurance company collects premiums from a number of participants, most of whom will typically receive less back than they put in. (Of course I'm simplifying here, such as when medical insurance pays for routine exams, the need for which is readily predictable.) In effect, we are paying the insurance company to take on some of what otherwise would be our risk.

We can think of an annuity as insurance in reverse, by which the insurance company or other financial institution pays us to take on some financial risk which we would otherwise not be subject to. This isn't the way annuities are marketed, of course, but a simple example will illustrate the point: An individual has some money in a U.S. bank account. The risk that this money will be lost is small, given that it's insured up to a limit by an agency of the US government. The account holder may then withdraw this money from the bank, and use it to purchase an annuity, which promises to pay some percentage of the underlying amount every year, starting several years from the time the annuity is purchased. Typically, this will be at a higher interest rate than that offered by the bank; otherwise the account holder would have no incentive to move the money. In buying the annuity, in return for this promise, the purchaser agrees to accept the risk that his lifespan will not be long enough for him to receive a reasonable return from the annuity. Under some unfortunate conditions, the return may in fact be zero, with all of the proceeds going to the insurance company, some of which it then uses to pay other annuity holders.

Annuities can bring large commissions to those who sell them to the public, leading to the meme that they are "sold" rather than "bought." In the area where I live there used to be a financial services company that exclusively sold annuities. I listened to their weekend radio broadcasts several times and did not once hear the term "annuity" used; the closest the hosts got was "product".

Can annuities be a valuable tool for financial planning? Sure; it's a maxim in the financial services industry that every financial product is the right product for somebody. But before you make a commitment you would be well-advised to know exactly what you're buying, why you're buying it, what it will cost you, what its limitations and risks are, what its alternatives are and how the proceeds from it will be taxed by federal and state governments. It will also be helpful to know whether the person proposing to sell you the annuity has the obligations of a fiduciary, and how that person will be compensated for the sale. Fortunately, the Internet has brought a degree of transparency to an industry that doesn't always like to be transparent. Do your homework: Compared to, say, investing in a stock mutual fund, if you decide that the annuity wasn't the right investment after all, it can be difficult and expensive to undo the decision and transfer the investment to an alternate vehicle.

Bonds

Bonds are in ways similar in behavior to annuities, so if you're thinking about buying an annuity you should also be looking at bonds. It's easiest to buy bonds through mutual funds, but it's also possible to buy them individually. In that case the interest rate and maturity should be fixed. ("Callable" bonds, though, are subject to being redeemed at the issuer's discretion.) Individual corporate bonds involve some risk of default, but that should not be a problem for US Treasury Bonds. You can buy individual US bonds through the Treasury Direct program (treasurydirect.gov).

Bonds can be less volatile than stocks but their values do move. When interest rates go up, so that the purchaser of a new bond can earn a higher interest than is offered by an equivalent existing bond, the existing bond's value decreases. (And vice versa, of course.) That's just one reason why bond prices can move.

Benchmarks

A benchmark is (typically) a market index selected to represent a risk profile similar to that of the mutual fund being measured. For example, for a mutual fund focusing on the largest US companies the Standard & Poor's (S&P) 500 Index would be an appropriate benchmark, whereas an international bond index would be inappropriate. Index funds seek to duplicate the performance of their selected benchmarks at low cost, while actively-managed funds seek to do better but generally at higher cost. To benchmark the performance of a portfolio that has investments in several funds, you can develop you own benchmark as a weighted average of the performance of a set of indices. The weights would represent your desired risk profile.

Should you use an adviser?

Investment adviser? Financial adviser? Fiduciary or not? Human or robot? There are many kinds of advisers and they don't necessarily offer the same services in the same way.

In the first few years of an engineer's career, the question will be largely moot, as he is unlikely to have enough assets for a typical human professional adviser to consider worthwhile. Some do specialize in small accounts and don't require a large minimum investment, and "robo-advisers" are reducing the cost of giving advice. Once the engineer has enough wealth to be interesting to an advisor, he might consider engaging one to professionally manage his investments. It's an option, not mandatory. Without doubt, as someone who got through engineering school you have the aptitude with numbers required to manage your own investments. The proof? I've met many financial advisers who started their careers as engineers. (Conversely, I have never met an engineer who used to be a financial adviser. Does that say something about our society, economy, and career choices?) However while you have the aptitude you don't necessarily have the knowledge, and as is also the case with our own profession, once acquired your knowledge must be continually refreshed. You may also not have the time or the interest. Fortunately, you shouldn't have to look too far to find someone willing to help you manage your money.

Your adviser can help manage both taxable and tax-deferred accounts, although in the case of the latter the adviser won't be able to directly manage money that is under the control of the employer's 401(k) custodian. On leaving a job, you can transfer money from the employer's 401(k) to an adviser-managed IRA. Just make sure that the money never goes directly to you, to avoid some nasty tax consequences. Your adviser can help you with the mechanics.

How do you pick an adviser? You've probably heard of several advisers already, as their professional ethics don't prevent them from advertising. In many areas, the airwaves especially on weekends are filled with infomercials from advisers pitching their approaches. If one sounds good, you should be able to arrange an introductory appointment without making a further commitment. Recommendations from friends can also be helpful. When you're interviewing an adviser find out whether she is ethically and legally required to act as a "fiduciary" or only required to recommend investments that are "suitable." A fiduciary is obligated to put her clients first. An adviser whose compensation is determined only as a percentage of the assets under her management ("fee-only") has her compensation aligned with your interests, and is probably a fiduciary. In contrast, an adviser who is paid by commission is incentivized at least in part to act to increase those commissions and so is probably not obligated to only use investments that are optimal for the client, merely appropriate. Such an adviser may use the term "fee-based", which is not the same thing as fee-only. A fee-based account may be appropriate for some investors, particularly those who wish to see active trading being done within their accounts. Active trading is simplest for tax-deferred accounts as frequent buying and selling may lead to complex tax returns in other kinds of accounts.

As a practical example of what it means to be a fiduciary, an adviser may know of three similar mutual funds that support her client's investment strategy in a similar way. As a fiduciary she would be required to choose the fund with the lowest cost (all else being equal); otherwise she would be free to select the fund that pays her the highest commission. The latter would still be a "suitable" investment.

If you use an adviser who is paid by commission, make sure you understand how the commission is paid. The commission may not be disclosed to you explicitly when the transaction is reported; it may instead be incorporated into the product price just as the fuel prices posted at gas stations don't necessarily break out taxes. Also, before you engage an adviser find out who the "custodian" is. The custodian is the entity that actually holds the money. Commonly, household-name financial companies like Fidelity act as custodians. Be wary of any adviser who also wants to also serve as custodian; the custodian should be completely independent of the adviser. If the proposed custodian is a company you've never heard of, check it out first.

What can you reasonably expect from your adviser? Advisers don't all offer the same set of services, but one basic service that almost all do offer is investment advice, although the term implies that it's up to the account owner to act on that advice, which isn't the typical case. Instead, ordinarily the adviser buys and sells securities in accordance with broad instructions and policies that the account owner agreed to in advance. These include the level of risk that the investor desires.

Let's assume a common arrangement in which the adviser earns an annual, all-in fee. We'll assume a fee rate of one percent of the account value per year. The fee may be assessed over several periods within a year, e.g., quarterly. Such an arrangement helps to align the adviser's compensation with your goals, but the adviser does get paid first – before you do. With the adviser taking one percent of the asset value per year, to break even (versus managing the money yourself the assets under her management would have to increase by one percentage point more. To evaluate this, you'll need to define a benchmark to reflect the risk level you've directed the adviser to take on. Either maintain a parallel (real or virtual self-managed portfolio with a similar risk profile, or use a simple market index or a combination of indices (e.g., stocks and bonds. Using similar risk profiles should minimize the creation of a perceived advantage or disadvantage for the adviser as a function of the stock market's short-term performance.

What if you get a benefit of one percentage point as a result of the adviser's work, so that you break even? The adviser will have earned a fee equal to one percent of your assets, and you will have earned exactly what you would have earned without the adviser's help. It's my belief, not readily provable, that most engineers with a little work could over time come within one percentage point of an adviser's results given the same risk tolerance. Ideally, then, in addition to managing your portfolio, with at most limited additional cost, the adviser should be able to help you with a variety of other financial matters. These could include financial projections, insurance advice, whether, when and where to purchase a home, support for estate planning (a task primarily directed by an estate-planning attorney), how to take Social Security benefits, and college financing.

You might, however, reasonably decide that you deserve a benefit equal in value to what the adviser gets, in which case the adviser should over time do two percentage points better than your benchmark – one percentage point for herself, and one for you. If the benchmark goes down one percent, the adviser's results should be up one percent. This amounts to the adviser doing perhaps 25% better than an unmanaged or minimally managed portfolio. I'm not sure that a lot of advisers can do this consistently over decades.

An adviser should be able to create a model of your financial position, and using the Monte Carlo method[186] give you a probability distribution that quantifies your chances of maintaining financial stability through your retirement. Such an analysis will either give you the comfort of knowing that you're likely to be OK, or give you reason to be concerned and act accordingly. However, a Monte-Carlo analysis isn't by itself a decision-support tool. For example, the program will likely make assumptions about your withdrawal strategy – as you draw down your retirement funds to meet living expenses, how do you tap into the varied investments in your portfolio? Do you emphasize tax-deferred investments, ordinary taxable accounts, or draw on them according to some rule? To answer questions like this, the adviser's model would have to iterate twice, running the entire set of Monte Carlo cases underneath a search strategy which modeled different decisions you might make concerning withdrawal strategies. That way you could determine an optimal withdrawal strategy given such constraints as the frequency with which you intend to replace automobiles.

You could pick a threshold – say a 95 percent chance of not outliving your savings – and then use the program to define an optimal financial strategy that neither excessively constrains your lifestyle nor creates undue risk of financial stress. Instead of several hundred cases, the analysis might have to consider millions. I am not aware of any financial adviser that does this, but there may be some. I acknowledge that many clients would have difficulty comprehending the output of such a study, but for engineers the results of such a study should be understandable, and for some, perhaps even entertaining.

Do remember, though, that "black swans", low-probability events like a pandemic that seriously damages the worldwide economy, can happen. Albert Einstein is said to have once asked his students how he could be alive: Given that lifetimes are finite and time is infinite, the probability that he was alive at the time should have been zero. Yet there he was. The resolution: You can't assign a probability when you already know the result.

Choose your adviser carefully. The retail financial industry is heavily regulated and the overwhelming majority of advisers are honest and ethical. Still, you have to keep your eyes open. Just two examples: In 2015 a national-scale financial company settled for $27.5 million a lawsuit by some of its own employees claiming among other things that the company breached its fiduciary duty by charging too much for its mutual funds.[187] And in 2012 advisor HSBC agreed to a $750,000 settlement with the Securities and Exchange Commission for allegedly misrepresenting its compensation practices.[188] (A settlement rarely if ever constitutes an admission of guilt. It may mean simply that the party offering the settlement does not wish to contest the issue in court.)

Financial management in retirement

Most of us are in the fortunate position of having to plan for many years of retirement once it starts. This means that the retirement planning process is nowhere close to being finished at the time you retire. There will likely be significant financial and tax consequences associated with the way you withdraw money from your retirement portfolio. You may be familiar with the common heuristic that up to four percent of your retirement savings can be withdrawn every year without placing your financial security in jeopardy. This can be a starting point, but there is much more to consider.

First, you have to think about risk, specifically the probability that you will outlive your money. To reduce that risk in the long term you may want to take on more risk in the short term, meaning keeping more of your investments in relatively volatile instruments, i.e., the stock market, than you might otherwise. Here a self-balanced portfolio can be better than a target-date fund, since you'll have more control over which investments your funds are withdrawn from in any given year. You can, for example, reduce withdrawals from equities and increase withdrawals from fixed-income instruments when you believe that prices of the former are at a low point and expect them to recover.

Second, think about inflation and compounding. The compound-interest tables that we've all seen show an astonishing accumulation of money even at even at a modest interest rate after enough years. I'll give you one example and leave the rest of the analysis to you and your spreadsheet: If you invest $1000 at the end of each year for 45 years at 5% interest, you'll have an account value of $159,700. If you wait five years to start, but otherwise stay with the same strategy, you'll have only $120,800 at the end of 45 years (40 years of investing) – a penalty of$38,900, or almost 25 percent, resulting from your not investing $5000 early on. But to account for 3% annual inflation, you'll have to divide the final amounts by 3.67. This results in present values of $43,498 for 45 years of investment versus $32,902 for 40 years.

Third, think about taxes. I've already discussed the fact that withdrawals from traditional 401(k)'s and IRA's are subject to taxation at ordinary-income rates. Some fraction of your Social Security benefits may also be subject to federal taxation, and depending on where you live they may be subject to state taxation as well. If you're primarily invested in traditional retirement accounts there may not be much you can do. However if some portion of your investable assets are in Roth accounts, or for that matter in non-tax-advantaged accounts, you can develop a strategy for payouts which uses just enough money from the latter sources to keep you from being put into a higher tax bracket. Here it will be helpful to be doing your taxes on commercial tax-preparation software, so you can simulate the effects of different decisions about how you take your money out. As tax laws often change, this technique should only be used for short-term planning.

Think also about how taxation is likely to evolve as you approach and go through retirement. Do you think taxes are more likely to go up or down? One rarely sees taxes go down, especially at the state and municipal level where borrowing is more difficult than at the federal level.

There isn't a one-size-fits-all optimal strategy here, but your financial adviser should be able to help you with your decisions.

Government programs in retirement

I've already discussed Social Security. Now we'll cover Medicare.

Medicare is a federal health insurance program for people age 65 and older, as well as for younger people with certain medical disabilities. The government makes available a great deal of information about the program; start at medicare.gov. You should begin learning about the details well before your 65th birthday. Here I'll give you a brief summary of the program as it stands at the time of writing and point out that there can be bad and long-lasting ramifications if you don't follow Medicare's rules to the letter. As with my other treatments of financial issues, this isn't intended or presented as a complete guide, but rather to motivate you to take the time to learn the details, of course getting help as you need it.

Medicare Part A covers inpatient care in hospitals, as well as certain other kinds of care. Most people sign up for it when they turn 65. Part A coverage is "free" for most workers in the sense that there is no monthly premium once they sign up for it, but people start paying for it with their first paychecks via a payroll deduction that's separate from

Social Security and has no maximum annual contribution. Coverage under Part A is funded by its own trust fund, the Hospital Insurance Trust Fund, which is predicted to be depleted by 2026.[189] Congress may act to increase revenue into the Fund in order to forestall such an occurrence.

Medicate Part B covers outpatient care, preventive care, and similar services. You'll have to pay a monthly premium for Part B. Premium costs rise periodically, and can also rise with income. Notwithstanding 40 or more years of Social Security payroll deductions and continuing Part B premiums, coverage under Part B is subject both to deductibles (you have to pay a specified amount before coverage starts) and coinsurance (Part B won't pay 100% of covered costs). So in planning for your retirement, make sure to account for unreimbursed medical costs.

Parts A and B together are called "Original Medicare".

Medicare Part C, also known as a Medicare Advantage Plan, is an optional health insurance policy issued by a private company. It subsumes Parts A and B and may add in certain other coverages. The Medicare Advantage Plan receives a monthly payment from Medicare on your behalf.

Medicare Supplemental Coverage, also sometimes called Medigap, is not Medicare at all but a health insurance policy from a private issuer that provides coverage for certain costs that are not covered by original Medicare. You pay for this coverage separately. You might have Medicare Advantage, a Medicare Supplement, or neither; but you would not have both a Medicare Advantage Plan and Medicare Supplemental Coverage.

Both Medicare Advantage Plans and Supplemental Coverage can reduce your out-of-pocket costs compared to Original Medicare, but of course both are paid programs whose cost you will need to include in your budgeting. (Some Medicare Advantage plans do not impose additional fees; they operate solely on the payments they receive from Medicare.) Some employers cover some of these costs for retirees and possibly their retired spouses. In considering either of these options make sure to understand not only costs but benefits; a plan that seems attractive initially can come with high out-of-pocket costs for products and services that are only partly covered.

Medicare Part D is optional coverage for prescription drugs. Part D coverage is provided by private companies and must be paid for separately. Many Medicare Advantage plans include prescription drug coverage that you wouldn't need to duplicate with Part D. In fact once you sign up for Medicare Advantage you may not be allowed to purchase Part D coverage even if the Advantage plan does not include a drug benefit. Part D brought us the infamous "doughnut hole" by which there is a gap between the initial coverage limit and the beginning of catastrophic coverage. Check current laws and your own health status to determine whether this might affect you.

Here's a hazard that can trip up people who aren't paying attention: The government wants to make sure that people don't wait to sign up for Part B until they really need it. Significant lifetime penalties in the form of increased premiums can be imposed on those who don't sign up once they become eligible. There may also be a delay in getting any Part B coverage at all. As long as you (and your non-working spouse) are covered through your current employment, Medicare will allow you to delay signing up for Part B. But then you must sign up within a Special Enrollment Period defined by the date at which your work-related coverage ended. The phrase "current employment" is used for good reason: For example, coverage under COBRA (The Consolidated Omnibus Budget Reconciliation Act of 1985), through which you can maintain your work-related medical coverage for a limited time after leaving your job, is not current employment. Therefore coverage under COBRA does not count in determining the Special Enrollment Period. You can be penalized for the rest of your life for waiting until your COBRA coverage ends before signing up for Part B. This fine print catches many unaware.[190]

Social Security, at least as of this writing, has an annual maximum contribution after which your earnings are no longer taxed. In contrast you must pay for Medicare throughout your working life with no limit on how much of your earnings are taxable. Taxation is through the Social Security system. One you're retired you'll keep paying – for uncovered medical costs and typically a Medicare Advantage program or Medicare Supplement plus drug coverage. In 2020, Fidelity Investments estimated that the average couple retiring that year could need $295,000 in assets (after tax, independent of long-term care costs) to pay unreimbursed medical costs over their lifetime.[191]

Dental, Vision and Hearing Care

No part of original Medicare covers routine dental, vision or hearing expenses. However, Medicare Advantage plans can offer these and other services. Military veterans may be able to get some services through the Department of Veterans Affairs. Your former employer may offer subsidized insurance coverage to retirees. Even without an employer subsidy, you might do well to get dental coverage under your employer's group rates, for up to 18 months under COBRA, once you've retired. There are also on the market dental insurance plans directed toward retirees, but do your homework.

Long-term Care

Medicare does not cover the cost of long-term care, which it defines as "…a range of services and support for your personal care needs. Most long-term care isn't medical care. Instead, most long-term care is help with basic personal tasks of everyday life like bathing, dressing, and using the bathroom, sometimes called activities of daily living." Many people purchase long-term care (LTC) insurance to help protect themselves against these costs, but the insurance itself can be expensive.

Medicaid

Medicaid is a program completely separate from Medicare. It is managed jointly by federal and state governments and is intended for people and families with low income. As an engineer you can only hope that you'll never need it during your working lifetime, but it does become important to some retirees. For the typical engineer it can be difficult to accumulate the resources, including long-term care insurance, owned assets, and pensions and annuities, to fully fund a sustained need for assisted living accommodation, residence in a nursing home, or intensive at-home help. Medicaid is your final backstop.

For workers who once earned respectable incomes but come to find their savings being drained away by massive living and care costs, entrance into Medicaid becomes an arms race between government rule-writers who want to see you truly impoverished before you can qualify, and specialized attorneys adept at finding ways for you to prevent the system from taking your last nickel or forcing you out of the home that you would rather be allowed to stay in. Medicaid law is so specialized that some lawyers in the field don't work on anything else – they would refer you to other attorneys for help with other elder-law issues such as drafting wills. In an earlier section I suggested that most engineers can learn to successfully manage their own investments. No such thing is true for elder law planning in general and Medicaid in particular; these are not places for a do-it-yourself approach. You need to have in place a comprehensive estate plan and then as necessary be prepared to enlist the help of an attorney specializing in Medicaid.

Final thoughts and the future of the profession

I hope that you've found this book both valuable and a good read. I wrote it in part because I hadn't seen anything quite like it, written from the viewpoint of a practitioner rather than an academic or executive. The vocation of engineering provides its members with both a livelihood and a creative outlet, but its members all too often experience salary compression, forced obsolescence, restrictions on their career mobility and having their advice ignored by their managers, the latter sometimes with disastrous results. At the same time, mechanisms that society uses to validate the qualifications of doctors, lawyers and the like are not applied consistently for us. This places engineering in an unusual position, with a unique combination of responsibilities and recognition. Perhaps the next generation of engineers will work to resolve it.

I believe that the defining challenge for the next several generations of American engineers, aside from keeping our country safe, will be to find ways to help maintain our first-world lifestyles with a much-diminished environmental footprint. Others who don't now enjoy these lifestyles would like to, and for the future of all, all will need to do so in a way that treads lightly. Notwithstanding climate change and resource depletion, we're not all going to voluntarily move into caves; there probably aren't enough available anyway. Engineering is going to have to move past onerous employment agreements, salary compression and age discrimination to provide the solutions our society will need. Will it?

David Deming, the Harvard researcher already mentioned in our discussion on salary compression, found that engineering jobs decreased relative to other STEM-related job from 2000 to 2012, while computer-oriented jobs increased.[192] However, government and industry have not stopped encouraging young people to study engineering. This does not bode well for the future of engineering, at a time where society is going to need what we can bring it.

In writing this book I was pleased to find that others had already asked some of the same questions I was asking, and reached many of the same conclusions. Their work is referenced throughout. However I was surprised to see that almost none of it was written by engineers or engineering faculty; economists and legal scholars predominate. Some professional societies are active in career issues but their names and publications rarely turned up in my Web searches. I hope we engineers haven't given up.

It could be that engineering in the United States is becoming a hybrid between technology and business. Without "soft skills" the engineer will be at a disadvantage. Does this mean that "hard" engineering skills will continue to migrate overseas? Four years of college are barely enough for us to learn what we have to learn now, and present salary structures do not encourage students to spend an extra year on campus to gain BA degrees in addition to their first degrees in engineering.

International competition will continue to be a factor in the American engineering world. As of this writing, a new Presidential administration has taken office and over time will demonstrate its views on issues like H-1B's and outsourcing. So far, there haven't been any major changes to the rules.

Like everyone else, we American engineers like to get paid. When we work at startups that don't yet have products we're generally paid with money supplied by investors. When we work on products and manufacturing processes f or existing companies our salaries compete with other corporate priorities, and when we design public works we compete for tax dollars. Availability of capital is a prerequisite for the success of the engineering profession. As I write this, with interest rates reaching record lows and stock market indices near record highs, availability of capital doesn't appear to be a problem. That may not always be the case, and let's recognize that trade deficits remove capital from our economy.

Medical economics concerns itself with the delivery of medical services, and legal economics looks at the economic consequences of aspects of the law. Engineering economics looks at the best ways to invest other peoples' money. This looks like just another way in which our profession occupies a challenging position.

Educational costs will probably keep inflating, but the COVID pandemic has forced a rethinking of how education is delivered. Educational institutions have learned how to deliver their services online, and students have learned to make use of them. Remote learning should be less expensive than the traditional model and make specialized education accessible to a wider audience. As of this writing it remains to be seen whether less-expensive options will remain available after the pandemic subsides, and whether cost savings attendant to these innovations will find their way to students. However, most engineering fields need some component of in-person learning; simulation is fine, but I believe that seeing a distillation column, a bridge or a nuclear reactor up close is better. We might end up with a hybrid, combining the cost saving associated with remote learning with on-campus instruction making use of specialized facilities.

Technical standards help ensure that compliant products are fit for their intended purposes, compatible with other products and systems with which they are to be used, and interchangeable with other compliant products. It would be difficult to argue against them, including international standards. Standards promote the interests of consumers by making it more difficult for one player in a market to lock out competitors; you can buy a USB memory flash drive with some confidence that it will work in your computer, no matter what brand name either of them carries. But standards can also serve to promote the interests of the party that sets them: Early in my career I worked with certain products that needed to comply with US standards. Standards used elsewhere were slightly more stringent, with the result that these overseas products could be used in the US but the US products had to be redesigned before they could be used overseas.

In the west in general, we're used to being the ones who are setting the standards. That may be changing now, to the detriment of American engineering. China has been using its Belt and Road initiative, by which it builds and maintains control over the infrastructure of developing nations, to impose industrial standards.[193] Patented Chinese technology is likely to help shrink the addressable market for western competition. And even where there is no patent protection, as Chinese standards become more prevalent western manufacturers may need to incur considerable expense to comply with them.

In defense and aerospace matters, counterfeit Chinese chips have crashed American military networks.[194] There were also allegations in 2018 that Chinese-manufactured hardware used widely in US server farms was compromised. This is a natural result of globalization of engineering and manufacturing.

In 2018 a jury in Madison, Wisconsin convicted Sinovel Wind Group Co. Ltd. of the Peoples Republic of China of stealing trade secrets from AMSC (formerly American Superconductor, Inc.) relating to controllers for wind turbines.[195] AMSC said that the theft resulted in the loss of almost 700 jobs and almost $1 billion in shareholder value.

The bipartisan Commission on the Theft of American Intellectual Property[196] in its 2019 review lauded the efforts of the US Government to elevate theft of intellectual property, especially by China, to the highest levels of government-to-government discussion. IP theft is, of course, also a threat to American engineering. Why should a company pay American engineers to develop technology only to have it stolen; it would be cheaper to just hire the people who are going to get it anyway.

Competition from China has been found to be more than a distant, diffuse threat to American manufacturing and its engineering community. A group of economists found that a reduction in US patent filings corresponded with China's entry into the World Trade Organization and the subsequent increase in imports from that country.[197] They attribute the change to a reduction in corporate R&D spending brought about by a decrease in US manufacturers' profitability.

With the United States so willing to educate so many from nations that are clearly hostile to us, should we not insist on some reciprocity? Shouldn't equal numbers of American students be allowed to study at, for example, China's most prestigious institutions of science and technology? This goes for other countries as well. Mandarin Chinese is by far *not* the most popular foreign language being studied in the United States, and opportunities for foreign language study in general appear to be diminishing. However funding sources may be available to help Americans study Chinese and other languages considered strategic to US needs.

Bibliography

There are many places where you can be trained to be an engineer. If you want to be educated about engineering, I recommend these three books:

The Revolt of the Engineers: Social Responsibility and the American Engineering Profession by Edwin T. Layton, originally published in 1971 and reprinted with a new preface in 1986. Layton discusses the efforts of American engineers to create and maintain an independent profession.

Falling Behind? Boom, Bust and the Global Race for Scientific Talent by Michael S. Teitelbaum, published in 2014. Teitelbaum describes the cyclical nature of the demand for engineers and scientists as well as the forces behind it.

Sold Out: How High-Tech Billionaires and Bipartisan Beltway Crapweasels are Screwing America's Best and Brightest Workers by Michelle Malkin and John Miano, published in 2016. Miano started out as a programmer and then went to law school to become an expert on immigration policy especially as it affects American technology workers. The writing has been described as "overheated", as the subtitle may suggest, but the book's assertions are supported by extensive research and citations.

Appendix A

Social Security Trends, 1995-2021

Year	FICA Max Contribution (1)	Consumer Price Index (2)	Social Security COLA (3)
1995	$61,200.00	147.8	2.60%
1996	$62,700.00	151.7	2.90%
1997	$65,400.00	156.3	4.20%
1998	$68,400.00	158.4	4.00%
1999	$72,600.00	161	2.50%
2000	$76,200.00	165.6	3.50%
2001	$80,400.00	171.7	2.60%
2002	$84,900.00	173.2	1.40%
2003	$87,000.00	177.7	2.10%
2004	$87,900.00	180.9	2.70%
2005	$90,000.00	186.3	4.10%
2006	$94,200.00	194	3.30%
2007	$97,500.00	197.559	2.30%
2008	$102,000.00	206.744	5.80%
2009	$106,800.00	205.7	0.00%
2010	$106,800.00	212.568	0.00%
2011	$106,800.00	216.4	1.70%
2012	$110,100.00	223.216	1.50%
2013	$113,700.00	226.52	1.70%
2014	$117,000.00	230.04	1.70%
2015	$118,500.00	228.294	0.00%
2016	$118,500.00	231.061	0.30%
2017	$127,200.00	236.854	2.00%
2018	$128,400.00	241.919	2.00%
2019	$132,900.00	245.133	1.60%
2020	$137,300.00	251.361	1.30%
2021	$142,800.00	255.296	5.90%

Notes to Appendix A

(1) https://www.ssa.gov/oact/cola/cbb.html accessed January 5, 2022

(2) CPI for Urban Wage Earners and Clerical Workers as of January of the specified year, https://www.ssa.gov/oact/STATS/cpiw.html, accessed January 5, 2022

(3) Adjustments are payable as of December of the specified year, but paid to recipients in January of the next year.
https://www.ssa.gov/oact/cola/colaseries.html, accessed January 5, 2022

Compound Annual Growth Rate for years 1995-2021 is 3.31 percent for FICA maximum contribution, 2.12% for Consumer Price Index

INDEX

[1] World Nuclear Association, "Fukushima Daiichi Accident", January 2021, https://www.world-nuclear.org/information-library/safety-and-security/safety-of-plants/fukushima-daiichi-accident.aspx, accessed March 5, 2021

[2] US Department of Labor, Bureau of Labor Statistics, Occupational Descriptions, https://www.bls.gov/ncs/ocs/ocs95apb.htm#engineer, accessed May 10, 2020

[3] American Society for Quality, Certification Catalog, https://asq.org/cert/catalog, accessed May 10, 2020

[4] Transportation Professional Certification Board, Inc., Home Page, http://www.tpcb.org, accessed May 10, 2020

[5] International Association for Radio, Telecommunications and Electromagnetics, "iNARTE EMC Design Engineer Certificate", https://inarte.org/certifications/inarte-emc-design-engineer-certification/, accessed November 2, 2020

[6] Texas Board of Professional Engineers and Land Surveyors, "Licensing Information", http://www.tbpe.state.tx.us/lic.htm, accessed May 11, 2020

[7] Walter W. Buchanan, "A Brief History of Engineering Technology and a Case for Applied Engineering", https://peer.asee.org/a-brief-history-of-engineering-technology-and-a-case-for-applied-engineering.pdf, accessed May 11, 2020

[8] Brandeis University, Office of Continuing Studies, http://www.brandeis.edu/continuing/one-program.php?prog_id=2, accessed June 24, 2020

[9] San Jose State University, Master of Science in Software Engineering, https://sjsu.edu/msse/, accessed June 24, 2020

[10] NCEES, PE Electrical & Computer Exam, https://ncees.org/pe-electrical-and-computer-exams-transition-in-2021/, accessed March 29, 2022.

[11] Wittliff, Dan, "Finally, a PE Path for Software Engineers" National Society of Professional Engineers, https://www.nspe.org/resources/pe-magazine/december-2012/finally-pe-path-software-engineers, January, 2011, accessed June 24, 2020

ENDNOTES

[12]National Council of Engineering Examiners, "NCEES discontinuing PE Software Engineering Exam", https://ncees.org/ncees-discontinuing-pe-software-engineering-exam/, posted March 13, 2018, accessed June 24, 2020

[13]U.S. National Science Foundation, "SESTAT: Characteristics of Scientists and Engineers in the United States", https://www.nsf.gov/statistics/us-workforce/, accessed June 24, 2020

[14]Somers, Darian and Josh Moody, "10 College Majors With the Best Starting Salaries", U.S. News and World Report, September 14, 2020, https://www.usnews.com/education/best-colleges/slideshows/10-college-majors-with-the-highest-starting-salaries?slide=12, accessed December 29, 2020

[15]Population Reference Bureau, "More U.S. Scientists and Engineers Are Foreign-Born",https://www.prb.org/usforeignbornstem/, accessed June 24, 2020

[16]Carnoy, Martin and Richard Rothstein, "What do international tests really show about U.S. student performance?", January 28, 2013, https://www.epi.org/publication/us-student-performance-testing/, accessed December 18, 2020

[17]U.S. Code of Federal Regulations, 29 CFR Part 541.301, https://www.govinfo.gov/content/pkg/CFR-2011-title29-vol3/pdf/CFR-2011-title29-vol3-part541.pdf, accessed June 25, 2020

[18]U.S. Code of Federal Regulations, 29 CFR 541.400, https://www.govinfo.gov/content/pkg/CFR-2011-title29-vol3/pdf/CFR-2011-title29-vol3-part541.pdf, accessed June 25, 2020

[19]National Council of Examiners for Engineering and Surveying®, "The History of the National Council of Examiners for Engineering and Surveying 1920-2004", http://ncees.org/wp-content/uploads/2012/11/The-History-of-NCEES-full.pdf, accessed June 25, 2020

[20]Renssalaer Polytechnic Institute, "The First Civil Engineers", http://www.lib.rpi.edu/archives/gallery/first_ce_grads/first_ce_grads.html, accessed June 25, 2020

[21]Roode, Benjamin, National Society of Professional Engineers,"The Federal PErspective", Jan/Feb 2011, https://www.nspe.org/resources/pe-magazine/february-2011/federal-perspective, accessed December 21, 2020

[22]Laws of New York State, Title 8, article 145, § 7208 (k), https://www.nysenate.gov/legislation/laws/EDN/7208, accessed March 31, 2022

[23]Umwelt Bundestamt (UBA, German Environmental Agency) "Lifetime of electrical appliances becoming shorter and shorter", https://www.umweltbundesamt.de/en/press/pressinformation/lifetime-of-electrical-appliances-becoming-shorter, accessed June 29, 2020

[24] Professional Engineers Ontario, "Report Unlicensed Individuals or Companies", https://www.peo.on.ca/public-protection/complaints-and-illegal-practice/report-unlicensed-individuals-or-companies, accessed March 31, 2022

[25]National Transportation Safety Board, "Safety Recommendation Report: Natural Gas Distribution System Project Development and Review (Urgent), November 14, 2018, https://www.ntsb.gov/investigations/AccidentReports/Reports/PSR1802.pdf

[26]Spinden, Paul M., "The Enigma of Engineering's Industrial Exemption to Licensure: The Exception that Swallowed a Profession", Liberty University School of Law, 2015, https://digitalcommons.liberty.edu/lusol_fac_pubs/72/, accessed October 30, 2020

[27] National Society of Professional Engineers, "Code of Ethics for Engineers", https://www.nspe.org/resources/ethics/code-ethics, accessed June 29, 2020

[28]National Society of Professional Engineers Position Statement No. 1766, "Engineering Faculty Licensure", https://www.nspe.org/resources/issues-and-advocacy/take-action/position-statements/engineering-faculty-licensure, accessed July 1, 2020

[29]U. S. Department of Education, "Accreditation in the United States", https://www2.ed.gov/admins/finaid/accred/accreditation.html, accessed July 1, 2020

[30]American Bar Association, "ABA Section of Legal Education and Admissions to the Bar Nominating Committee Announces 2020-2021Council Slate", https://www.americanbar.org/content/dam/aba/administrative/legal_ed ucation_and_admissions_to_the_bar/20-21-legal-ed-council-slate-announcement.pdf, accessed July 2, 2020

[31] Accreditation Board for Engineering and Technology, Engineering Accreditation Commission, EAC Executive Committee, https://www.abet.org/about-abet/governance/accreditation-commissions/, accessed July 13, 2020

[32] Federal Communications Commission, Leadership, https://www.fcc.gov/about/leadership, accessed January 11, 2021

[33]Anon., "Trump science nominees have fewer advanced degrees in the field", CBS News, December 5, 2017, https://www.cbsnews.com/news/trump-science-nominees-missing-advanced-science-degrees/, accessed November 4, 2020

[34]United States Army, "Direct Commission Officers", https://www.goarmy.com/careers-and-jobs/become-an-officer/how-to-become-an-officer-in-the-army/direct-commission.html, accessed July 21, 2020

[35]Center for Responsive Politics, "Client Profile: American Medical Assn", October 23, 2020, https://www.opensecrets.org/federal-lobbying/clients/summary?cycle=2020&id=D000000068, accessed January 11, 2021

[36]Center for Responsive Politics, "Client Profile: IEEE-USA", October 23, 2020, https://www.opensecrets.org/federal-lobbying/clients/summary?id=D000079271, accessed March 31, 2022

[37] National Aeronautics and Space Administration, "Report to the President by the Presidential Commission on the Space Shuttle Challenger Accident", Chapter 5, June 6, 1986, accessed March 31, 2022, https://history.nasa.gov/rogersrep/v1ch5.htm

[38] U.S. House of Representatives, Majority Staff of the Committee on Transportation and Infrastructure, "The Design, Development & Certification of the Boeing 737 MAX", September 2020, https://transportation.house.gov/imo/media/doc/2020.09.15%20FINAL%20737%20MAX%20Report%20for%20Public%20Release.pdf, accessed September 24, 2020

[39] Shepardson, David, "Honda confirms 17th U.S. death in Takata air bag rupture", Reuters, October 4, 2020, https://www.reuters.com/article/honda-takata/honda-confirms-17th-u-s-death-in-takata-air-bag-rupture-idUSKBN26P04K, accessed November 3, 2020

[40] Anon., "Takata whistleblower says air bag warning was "ethical duty", Kyodo News, April 28, 2018, https://english.kyodonews.net/news/2018/04/257816b66f02-takata-whistleblower-says-air-bag-warning-was-ethical-duty.html, accessed November 3, 2020

[41] United States Department of Justice, Office of Public Affairs, "Volkswagen Engineer Pleads Guilty for His Role in Conspiracy to Cheat U.S. Emissions Tests", September 9, 2016, https://www.justice.gov/opa/pr/volkswagen-engineer-pleads-guilty-his-role-conspiracy-cheat-us-emissions-tests, accessed September 25, 2020

[42] United States Department of Justice, Office of Public Affairs, "Volkswagen Engineer Sentenced for His Role in Conspiracy to Cheat U.S. Emissions Tests", August 25, 2017, https://www.justice.gov/opa/pr/volkswagen-engineer-sentenced-his-role-conspiracy-cheat-us-emissions-tests, accessed September 25, 2020

[43] National Society of Professional Engineers, "NSPE Ethics in Employment Task Force Report", https://web.archive.org/web/20060927020844/http://www.nspe.org/ethics/eh1-report.asp, accessed July 21, 2020

[44] University of Massachusetts, News & Media Relations "Most Engineering Schools Do Not Require a Course in Ethics, UMass Amherst Professor Finds", March 6, 2000, accessed March 31, 2022 https://www.umass.edu/archivenewsoffice/article/most-engineering-schools-do-not-require-course-ethics-umass-amherst-professor-finds

ENDNOTES

[45] Accreditation Board for Engineering and Technology, "Criteria for Accrediting Engineering Programs, 2016-2017", https://www.abet.org/accreditation/accreditation-criteria/criteria-for-accrediting-engineering-programs-2016-2017/, accessed July 21, 2020

[46] Musselman, Craig N.; J.D. Nelson and M.L. Philips, "A Primer on Engineering Licensure in the United States ", https://www.nspe.org/sites/default/files/resources/pdfs/blog/ASEE-A-Primer-on-Engineering-Licensure-in-the-United-States.pdf, accessed July 22, 2020

[47] The Carnegie Classification of Institutions of Higher Education, Home Page, http://carnegieclassifications.iu.edu/, accessed July 28, 2020

[48] Dale, Stacy and Alan B. Krueger, "Estimating the Return to College Selectivity ofer the Career Using Administrative Earnings Data", National Bureau of Economic Research, June, 2011, https://www.nber.org/system/files/working_papers/w17159/w17159.pdf, accessed October 29, 2020

[49] Benderly, Beryl Leiff "Danger in School Labs: Accidents Haunt Experimental Science", Scientific American, https://www.scientificamerican.com/article/danger-in-school-labs/ accessed July 28, 2020

[50] Bass, Carole, "After a fatal accident, new safety standards", Yale Alumni Magazine, November/December 2011, https://yalealumnimagazine.com/articles/3288-after-a-fatal-accident-new-safety-standards, accessed July 28, 2020

[51] Mills, Julie; Mary Ayre, David Hands, Pam Cardin, "Learning About Learning Styles: Can This Improve Engineering Education?", https://pdfs.semanticscholar.org/a0bb/94a915c99689795e79109bfc62d27be1b7f0.pdf, accessed July 20, 2020

[52] North Carolina State University, College of Engineering, "Richard Felder's Legacy Website", https://www.engr.ncsu.edu/stem-resources/legacy-site/. accessed July 29, 2020

[53]American Association of University Professors, "Background Facts on Contingent Faculty",
https://www.aaup.org/issues/contingency/background-facts, accessed July 20, 2020

[54]Donaldson, Krista, Gary Lichtenstein, and Sheri Sheppard, "Socioeconomic Status and the Undergraduate
Engineering Experience: Preliminary Findings from Four American Universities", Center for the Advancement of Engineering Education, http://depts.washington.edu/celtweb/caee/CAEE_Briefs_PDFs/Socioecono micStatus_Donaldson_ASEE08.pdf, accessed July 29, 2020

[55] U.S. Department of Veterans Affairs, "Edith Nourse Rogers STEM Scholarship", https://www.va.gov/education/other-va-education-benefits/stem-scholarship/, October 26, 2021, accessed January 5, 2022

[56]Cooper, Preston "How How Unlimited Student Loans Drive Up Tuition", Forbes Magazine,
https://www.forbes.com/sites/prestoncooper2/2017/02/22/how-unlimited-student-loans-drive-up-tuition/#4c88a3352b63, Feb. 22. 2017, accessed July 29, 2020

[57]Simon, Caroline, "Bureaucrats And Buildings: The Case For Why College Is So Expensive", Forbes Magazine,
https://www.forbes.com/sites/carolinesimon/2017/09/05/bureaucrats-and-buildings-the-case-for-why-college-is-so-expensive/#2d9a8671456a, Sept. 5, 2017, accessed October 28, 2020

[58] Chingos, Matthew, "End government profits on student loans: Shift risk and lower interest rates", Brookings Institution,
https://www.brookings.edu/research/end-government-profits-on-student-loans-shift-risk-and-lower-interest-rates/, April 30, 2015, accessed July 29, 2020

[59]U.S. Department of Education, Public Service Loan Forgiveness, https://studentaid.gov/manage-loans/forgiveness-cancellation/public-service, accessed July 29, 2020

[60]American Society of Civil Engineers, Civil Engineering Body of Knowledge for the 21st Century, Second Edition,
https://sp360.asce.org/PersonifyEbusiness/Merchandise/Product-Details/productId/263831792, 2008, accessed March 31, 2022

[61]"NCEES adopts position statement on the future of engineering licensure". https://ncees.org/ncees-adopts-position-statement-future-engineering-licensure/, August 1, 2015, accessed July 29, 2020

[62] NCEES Position Statement 35, Future Education Requirements for Engineering Licensure, https://ncees.org/about/publications/ncees-position-statement-35/, 2017, accessed July 29, 2020

[63]American Physical Therapy Association, "Physical Therapist (PT) Education Overview", http://www.apta.org/For_Prospective_Students/PT_Education/Physical_Therapist_(PT)_Education_Overview.aspx, accessed July 29, 2020

[64]American Physical Therapy Association, "Physical Therapist Assistant (PTA) Education Overview", http://www.apta.org/PTAEducation/Overview/, accessed July 29, 2020

[65]American Physical Therapy Association, "Guide to Becoming a Physical Therapy Assistant or Aide", https://www.physicaltherapyaide.org/, accessed July 29, 2020

[66]International Association for Continuing Education and Training, Home Page, https://www.iacet.org, accessed July 30, 2020

[67]Boston University, "LEAP: A Masters Program for Non-Engineers". http://www.bu.edu/eng/prospective-graduate/leap/, accessed July 30, 2020

[68]Kay, Grace "Will STEM Degrees save the MBA?", Forbes Magazine, June 17, 2019, https://www.forbes.com/sites/gracekay/2019/06/17/will-stem-degrees-save-the-mba/#614f2cd667c7, accessed July 30, 2020

[69]US Citizenship and Immigration Service, "Optional Practical Training (OPT) for F-1 Students", https://www.uscis.gov/working-in-the-united-states/students-and-exchange-visitors/optional-practical-training-opt-for-f-1-students, accessed July 30, 2020

[70] National Professional Science Master's Association (NPSMA), Home Page, https://www.professionalsciencemasters.org/, accessed July 30, 2020

[71]"Education: Engineer Shortage" Time Magazine, April 21, 1952, https://content.time.com/time/subscriber/article/0,33009,889495,00.html, accessed March 31, 2022

[72] U.S. Department of Labor, Service Contract Act of 1965, as Amended, https://www.dol.gov/sites/dolgov/files/WHD/legacy/files/serv01.pdf, accessed April 16, 2020

[73] Comptroller General of the United States, "Special Procurement Procedures Helped Prevent Wage Busting Under Federal Service Contracts in the Cape Canaveral Area", https://www.gao.gov/assets/130/121756.pdf, accessed April 16, 2020

[74] Teitelbaum, Michael S., "The Myth of the Science and Engineering Shortage", March 19, 2014, The Atlantic, https://www.theatlantic.com/education/archive/2014/03/the-myth-of-the-science-and-engineering-shortage/284359/, accessed November 2, 2020

[75] Paliwal, Dinesh, "Engineering Is The New Liberal Arts" Feb.24, 2016, Forbes Magazine, https://www.forbes.com/sites/currentaccounts/2016/02/24/engineering-is-the-new-liberal-arts/?sh=583bcac87ac8, accessed November 2, 2020

[76] U.S. Government Publishing Office, Public Law 414, https://www.govinfo.gov/content/pkg/STATUTE-66/pdf/STATUTE-66-Pg163.pdf, accessed August 2, 2020

[77] United States Department of Labor, Wage and Hour Division, "H-1B Program", https://www.dol.gov/agencies/whd/immigration/h1b, accessed August 2, 2020

[78] U.S. Government Printing Office, Public Law 106-313, https://www.govinfo.gov/content/pkg/PLAW-106publ313/pdf/PLAW-106publ313.pdf, accessed August 2, 2020

[79] Wikipedia, "H-1B Applications Approved", https://en.wikipedia.org/wiki/H-1B_visa, accessed August 3, 2020

[80] BCIS, "Initial H-1B Beneficiaries by occupation and region and country of birth: fiscal year 2000" http://www.bcis.gov/graphics/shared/aboutus/statistics/00yrbk_Temp/TempExclTables/Table44.xls, information no longer electronically retrievable

[81] US National Science Foundation, "Science and Engineering Indicators" https://ncses.nsf.gov/pubs/nsb20197/trends-in-undergraduate-and-graduate-s-e-degree-awards#figureCtr711, accessed August 3, 2020

[82]Ibid.

[83]US Citizenship and Immigration Service, "Characteristics of H-1B Specialty Occupation Workers, Fiscal Year 2019 Annual Report to Congress", March 5, 2020, https://www.uscis.gov/sites/default/files/document/reports/Characteristics_of_Specialty_Occupation_Workers_H-1B_Fiscal_Year_2019.pdf, accessed August 3, 2020

[84]United States Department of Labor, Employment & Training Administration, "Prevailing Wage Information and Resources"; January 15, 2009, https://www.dol.gov/agencies/eta/foreign-labor/wages, accessed March 31, 2022

[85]United States Government Accountability Office, "H-1B Visa Program: Reforms Are Needed to Minimize the Risks and Costs of Current Program", January 14, 2011, https://www.gao.gov/products/GAO-11-26, accessed August 3, 2020

[86]United States Government Accountability Office, "H-1B Visa Program: Reforms Are Needed to Minimize the Risks and Costs of Current Program", January 2011, https://www.gao.gov/assets/320/314501.pdf, accessed August 3, 2020

[87] United States Department of Labor, User Guide for filling out the ETA-9035, https://flag.dol.gov/sites/default/files/2019-09/OFLC_Application_Modernization_How_To_Submit_ETA-9035_9035E_Application_User_Guide.pdf , accessed August 4, 2020

[88]United States Government, Federal Register, "Labor Condition Applications and Requirements for Employers Using Nonimmigrants on H-1B Visas in Specialty Occupations and as Fashion Models; Labor Certification Process for Permanent Employment of Aliens in the United States", December 20, 2000, https://www.federalregister.gov/documents/2000/12/20/00-32088/labor-condition-applications-and-requirements-for-employers-using-nonimmigrants-on-h-1b-visas-in, accessed August 4, 2020

[89] United States Citizenship and Immigration Service, "H-1B Approvals for the top 30 employers with the most continuing and initial approvals, fiscal year 2018", https://www.uscis.gov/sites/default/files/document/data/Top-30-H-1B-Employer-Table-FY-2018.pdf, accessed August 5, 2020

[90] United States Securities and Exchange Commission, "Cognizant Technology Solutions Corp SEC CIK #0001058290", https://sec.report/CIK/0001058290, accessed August 5, 2020

[91] Tata Consultancy Services Ltd., Press Release, "TCS Delivers Steady Growth in Q2", October 10, 2019, https://www.tcs.com/content/dam/tcs/investor-relations/financial-statements/2019-20/q2/IND%20AS/Press%20Release%20-%20INR.pdf, accessed August 5, 2020

[92] Infosys, Ltd., "About Us", History, https://www.infosys.com/about/history.html, accessed August 6, 2020

[93] Bloomberg News (company profiles), "Wipro Ltd.", https://www.bloomberg.com/profile/company/WPRO:IN, accessed August 6, 2020

[94] United States Code, 8 USC 1184: Admission of nonimmigrants, August 5, 2020, https://uscode.house.gov/view.xhtml?req=granuleid%3AUSC-prelim-title8-section1184&num=0&edition=prelim, accessed August 6, 2020

[95] United States House of Representatives, Office of the Clerk, "Lobbying Disclosure", July 2020, https://lobbyingdisclosure.house.gov/, accessed August 7, 2020

[96] United States House of Representatives, Office of the Clerk, Lobbying Report, https://disclosurespreview.house.gov/ld/ldxmlrelease/2020/Q1/301174065.xml, accessed August 7, 2020.

[97] Ruark, Eric, "Immigration Lobbying: A Window into the World of Special Interests", Federation for American Immigration Reform, https://www.fairus.org/sites/default/files/2017-08/fair_lobbying_report.pdf?docID=2421

[98]Broache, Annie, "Gates calls for "infinite" H-1B's, better schools", Cnet, March 8, 2007, https://www.cnet.com/news/gates-calls-for-infinite-h-1bs-better-schools/, accessed August 10, 2020

[99]Thibodeau, Patrick, "Bloomberg champions limitless H-1B visas", Computerworld, January 25, 2016, https://www.computerworld.com/article/3026241/bloomberg-champions-limitless-h-1b-visas.html, accessed August 10, 2020

[100]American Immigration Lawyers Association, "About", https://www.aila.org/about, accessed August 10,2020

[101]Preston, Julia "Pink slips at Disney. But First, Training Foreign Replacements", New York Times, June 3, 2015, https://www.nytimes.com/2015/06/04/us/last-task-after-layoff-at-disney-train-foreign-replacements.html, accessed August 10, 2020

[102]Thibodeau, Patrick, "Southern California Edison IT Workers 'beyond furious' over H-1B replacements", Computerworld, Feb. 4, 2015, https://www.computerworld.com/article/2879083/southern-california-edison-it-workers-beyond-furious-over-h-1b-replacements.html, accessed July 16, 2021

[103]Rojas, Leslie Berenstein, Southern California Public Radio, "Feds find no wrongdoing by firm that provided foreign workers to Edison", October 22, 2015, https://www.scpr.org/news/2015/10/22/55178/feds-conclude-investigation-into-firm-that-provide/, accessed August 10, 2020

[104]Thibodeau, Patrick, Computerworld, "Laid-off IT workers muzzled as H-1B debate heats up", January 28, 2016, https://www.computerworld.com/article/3027640/laid-off-it-workers-muzzled-as-h-1b-debate-heats-up.html, accessed August 10, 2020

[105]U.S. Citizenship and Immigration Services, "Combating Fraud and Abuse in the H-1B Visa Program", February 9, 2021, https://www.uscis.gov/scams-fraud-and-misconduct/report-fraud/combating-fraud-and-abuse-in-the-h-1b-visa-program, accessed July 16, 2021

ENDNOTES

[106] United States Department of Justice, "Justice Department Files Lawsuit Against Facebook for Discriminating Against U.S. Workers", December 3, 2020, https://www.justice.gov/opa/pr/justice-department-files-lawsuit-against-facebook-discriminating-against-us-workers, accessed December 15, 2020

[107] United States Department of Justice, "Justice, Labor Departments Reach Settlements with Facebook Resolving Claims of Discrimination Against U.S. Workers and Potential Regulatory Recruitment Violations", October 19, 2021, https://www.justice.gov/opa/pr/justice-labor-departments-reach-settlements-facebook-resolving-claims-discrimination-against, accessed January 4, 2022

[108] United States Department of Homeland Security, Citizenship and Immigration Services, "O-1 Visa: Individuals with Extraordinary Ability or Achievement, May 29, 2020, https://www.uscis.gov/working-in-the-united-states/temporary-workers/o-1-visa-individuals-with-extraordinary-ability-or-achievement, accessed August 10, 2020

[109] Anderson, Stuart, "55% Of America's Billion-Dollar Startups Have An Immigrant Founder", Forbes Magazine, October 25, 2018, https://www.forbes.com/sites/stuartanderson/2018/10/25/55-of-americas-billion-dollar-startups-have-immigrant-founder/, accessed August 10,2020

[110] "Combating Fraud and Abuse in the H-1B Visa Program", op. cit.

[111] United States Citizenship and Information Services, "H-1B Petitions by Gender and Country of Birth, Fiscal Year 2018", October 5, 2018, https://www.uscis.gov/sites/default/files/document/reports/h-1b-petitions-by-gender-country-of-birth-fy2018.pdf, accessed August 11, 2020

[112] Associated Press, "Former arms company engineer pleads guilty to weapons count", February 15, 2020, https://apnews.com/ed9f3301f016d9ab9bf500ea0591afe3, accessed August 11, 2020

[113] Spalding, Robert S. (Brig. Gen. USAF, retired) and A.J. Rice, "How US visas can help China's intel agencies to spy on us", The Hill, April 13, 2020, https://thehill.com/opinion/international/491947-how-us-visas-can-help-chinas-intel-agencies-to-spy-on-us, accessed August 11, 2020

ENDNOTES

[114]O'Keefe, Kate and Aruna Viswanatha, "Chinese Diplomats Helped Military Scholars Visiting the U.S. Evade FBI Scrutiny, U.S. Says", Wall Street Journal, August 25, 2020, https://www.wsj.com/articles/chinese-diplomats-helped-visiting-military-scholars-in-the-u-s-evade-fbi-scrutiny-u-s-says-11598379136, accessed December 15, 2020

[115] United States Environmental Protection Agency, "What Is Emissions Trading?", https://www.epa.gov/emissions-trading-resources/what-emissions-trading, accessed April 22, 2020,

[116] United States Department of State, Report of the Visa Office 2019, Table XVI(A), Classes of Nonimmigrants Issued Visas (including Border Crossing Cards) Fiscal Years 2015-2019, https://travel.state.gov/content/dam/visas/Statistics/AnnualReports/FY2019AnnualReport/FY19AnnualReport-TableXVI-A.pdf, accessed August 17, 2020

[117]Ibid.

[118]American Society of Plumbing Engineers, Membership & Global Community, https://www.aspe.org/membership-global-community/, accessed August 20, 2020

[119] National Society of Professional Engineers, www.nspe.org, accessed April 4, 2022

[120]American Association of Engineering Societies, http://www.aaes.org. Website available on August 25, 2020

[121]US National Labor Relations Board, "National Labor Relations Act", https://www.nlrb.gov/guidance/key-reference-materials/national-labor-relations-act, accessed February 23, 2021

[122]US Bureau of Labor Statistics, Press Release "Union Members Summary", January 22, 2020, https://www.bls.gov/news.release/union2.nr0.htm, accessed April 4, 2022

[123] Society of Professional Engineering Employees in Aerospace, Home Page, https://www.speea.org/, accessed April 4, 2022

[124] International Federation of Professional and Technical Engineers, Home page, http://www.ifpte.org

[125] International Federation of Professional and Technical Engineers, "IFPTE Commends Sen. Durbin and Sen. Grassley for Introducing the Bipartisan H-1B and L-1 Visa Reform Act", https://www.ifpte.org/news/ifpte-commends-sen-durbin-and-sen-grassley-for-introducing-the-bipartisan-h-1b-and-l-1-visa-reform-act, accessed April 4, 2022

[126] Meredith, John W. and Pender M. McCarter, "History of IEEE-USA: 1973-2009 – An Overview of Four Decades", https://ieeeusa.org/assets/about/history/IEEE-USA-History-1973-2009.pdf, accessed April 4, 2022

[127] National Society of Professional Engineers, "Engineer Membership in Labor Union", Case 62-5, https://www.nspe.org/resources/ethics/ethics-resources/board-ethical-review-cases/engineer-membership-labor-union, accessed August 26, 2020

[128] Department for Professional Employees, AFL-CIO, https://www.dpeaflcio.org/, accessed August 26, 2020

[129] Greenhouse, Steven, "TECHNOLOGY; Temp Workers At Microsoft Win Lawsuit", New York Times, December 13, 2000, https://www.nytimes.com/2000/12/13/business/technology-temp-workers-at-microsoft-win-lawsuit.html, accessed August 26, 2020

[130] U S. Internal Revenue Service, Publication 15A Section 2, "Employee or Independent Contractor?", https://www.irs.gov/pub/irs-pdf/p15a.pdf, accessed August 26, 2020

[131] State of California, "AB-5 Worker status: employees and independent contractors.(2019-2020)", https://leginfo.legislature.ca.gov/faces/billTextClient.xhtml?bill_id=201920200AB5, accessed February 28, 2021

[132] Shoot, Brittany, "Amazon's New $15 Minimum Wage Jobs Will Lose Bonuses and Stock Option Awards", October 3, 2018, Fortune Magazine, http://fortune.com/2018/10/03/amazon-warehouse-workers-minimum-wage-incentives-stock-options/, accessed August 28, 2020

[133] Deming, David, "Engineers Start Fast, but Poets Can Catch Up", New York Times, September 29, 2019. A similar article (not identical) appears at https://www.nytimes.com/2019/09/20/business/liberal-arts-stem-salaries.html, accessed September 25, 2020

[134]Deming, David and Kadeem Noray, "STEM Careers and the Changing Skill Requirements of Work", June 20, 2019, https://oconnell.fas.harvard.edu/files/ddeming/files/dn_stem_june2019.pdf, accessed October 27, 2020

[135]Miller, Steven, "Address Pay Compression or Risk Employee Flight", Society for Human Resource Management, June 1, 2018, https://www.shrm.org/resourcesandtools/hr-topics/compensation/pages/address-pay-compression-or-risk-employee-flight.aspx, accessed August 31, 2020. The full report is available for purchase. See Pearl Meyer, "Salary Compression Practices in the United States", October 2017, https://www.pearlmeyer.com/knowledge-share/research-report/salary-compression-practices-in-the-united-states, accessed August 31, 2020

[136]American Society of Civil Engineers, "ASCE Guidelines for Engineering Grades", https://www.asce.org/-/media/asce-images-and-files/career-and-growth/early-career-engineer/engineering-grades-brochure.pdf, accessed August 31, 2020

[137]Kaufman, Wendy, "The Few, the Tech-Savvy Few: Option Millionaires", National Public Radio, February 11, 2007, http://www.npr.org/templates/story/story.php?storyId=7324048, accessed August 31, 2020

[138]Brumberg, Bruce, "Restricted Stock, Restricted Stock Units, Restricted Securities: Understanding The Differences And Why They Matter", August 31, 2020, Forbes, https://www.forbes.com/sites/brucebrumberg/2020/08/31/restricted-stock-restricted-stock-units-restricted-securities-understanding-the-differences-and-why-they-matter/, accessed November 30, 2020

[139]Cussen, Mark P., "An Introduction to Incentive Stock Options", August 10, 2021, https://www.investopedia.com/articles/stocks/12/introduction-incentive-stock-options.asp, accessed April 4, 2022

[140]Davidson, Adam, "Why are Harvard Graduates in the Mailroom", New York Times Magazine, Feb. 26, 2012, https://www.nytimes.com/2012/02/26/magazine/why-are-harvard-graduates-in-the-mailroom.html, accessed February 26, 2012

[141]Fowler, Susan, "Reflecting on one very, very strange year at Uber", February 19, 2017, https://www.susanjfowler.com/blog/2017/2/19/reflecting-on-one-very-strange-year-at-uber, accessed September 22, 2020

[142]Society for Human Resources Management, "Code of Ethics", Amended November 21, 2014, https://shrm.org/about-shrm/Pages/code-of-ethics.aspx, accessed September 22, 2020

[143]Justia US Law, "Reddy v. Community Health Foundation of Man", https://law.justia.com/cases/west-virginia/supreme-court/1982/15453-4.html, accessed March 9, 2021

[144]"Bar Code Designer Defied Instructions", obituary of George Laurer (1925-2019), Wall Street Journal, December 14/15, 2019

[145] Yale University, Office of Cooperative Research, Yale University Patent Policy, February 23, 1998,https://ocr.yale.edu/policies accessed July 16, 2021

[146]Starr, Evan P, Norman Bishara and J.J. Prescott, "Noncompetes in the U.S. Labor Force", May 7, 2020; revised May 18, 2020. Journal of Law and Economics, forthcoming. Available at SSRN: https://ssrn.com/abstract=2625714, accessed march 23, 2021

[147]Dougherty, Conor "How Noncompete Clauses Keep Workers Locked In", The New York Times, May 13, 2017, https://www.nytimes.com/2017/05/13/business/noncompete-clauses.html, accessed September 1, 2020

[148]Starr, e. al., Op. cit.

[149]Starr, e. al., Op. cit.

[150]Starr, e. al., Op. cit.

[151]The Disciplinary Board of the Supreme Court of Pennsylvania, "Pennsylvania Rule of Professional Conduct (includes amendments through September 14, 2019)",

https://www.padisciplinaryboard.org/Storage/media/pdfs/20191218/203914-rpc2019-09-14.pdf, accessed September 1, 2020

ENDNOTES

[152]Massachusetts General Laws, Part 1, Title XVI, Chapter 112, https://malegislature.gov/Laws/GeneralLaws/PartI/TitleXVI/Chapter112/Section12X, accessed September 1, 2020

[153]Illinois General Assembly, Illinois Complied Statues, (820 ILCS 17/) Broadcast Free Market Act, https://www.ilga.gov/legislation/ilcs/ilcs3.asp?ActID=2390&ChapterID=68, accessed September 1, 2020

[154]American Federation of Television and Radio Artists, "AFTRA Successful in Banning Non-Compete Clauses for Broadcasters in Illinois", AFTRA Magazine, Winter 2002

[155]California Business and Professions Code, Section 16600 – 16607, https://leginfo.legislature.ca.gov/faces/codes_displayText.xhtml?lawCode=BPC&division=7.&title=&part=2.&chapter=1.&article, accessed September 1, 2020

[156]Starr, e. al., Op. cit.

[157]Dougherty, Conor "How Noncompete Clauses Keep Workers Locked In", The New York Times, May 13, 2017, https://www.nytimes.com/2017/05/13/business/noncompete-clauses.html, accessed September 1, 2020

[158] United States Department of Justice, complaint filed in the United States District Court for the District of Columbia, https://www.justice.gov/atr/case-document/file/483451/download, accessed September 1, 2020

[159]United States Department of Justice, Final Judgment, U.S. v. Adobe Systems, Inc., et al., March 17, 2011, https://www.justice.gov/atr/case-document/final-judgment-0, accessed September 1, 2020

[160]"Order granting plaintiffs' supplemental motion for class certification, Case5:11-cv-02509-LHK", posted by law firm Lief, Cabraser, Heimann & Bernstein, https://www.lieffcabraser.com/pdf/high-tech-antitrust-class-certification-order.pdf, accessed September 1, 2020

[161]"Order granting plaintiffs' motion for final approval of class action settlement with defendants adobe systems incorporated, apple inc., google inc., and intel corporation, Case5:11-cv-02509-LHK", posted by law firm Lief, Cabraser, Heimann & Bernstein, https://www.lieffcabraser.com/pdf/high_tech_final_order.pdf, accessed September 1, 2020

[162]"High-Tech Employee Antitrust Settlement", Gilardi & Co. LLC, http://www.hightechemployeelawsuit.com/, accessed September 1, 2020

[163]"Disney Settles Anti-Poaching Lawsuit", Forbes Magazine, February 2, 2017, https://www.forbes.com/sites/legalentertainment/2017/02/02/disney-settles-anti-poaching-lawsuit/, accessed September 1, 2020

[164]Animation Workers Antitrust Settlements Website, Sept. 20, 2018, www.animationlawsuit.com, accessed July 16, 2021

[165]Legal Information Institute, "Sherman Antitrust Act", https://www.law.cornell.edu/wex/sherman_antitrust_act, accessed September 2, 2020

[166]Lobel, Orly "NDAs Are Out of Control. Here's What Needs to Change" Harvard Business Review, January 30, 2018, https://hbr.org/2018/01/ndas-are-out-of-control-heres-what-needs-to-change, accessed October 7, 2020

[167]Mabud, Rakeen, "Google Put An End To Forced Arbitration -- And Why That's So Important", Forbes Magazine, February 26, 2019, https://www.forbes.com/sites/rakeenmabud/2019/02/26/worker-organizing-results-in-big-change-at-google/, accessed September 2, 2020

[168]Newman, Romy "why-your-company-needs-to-offer-unlimited-vacation", Inc. (magazine), April 10, 2017, https://www.inc.com/romy-newman/why-your-company-needs-to-offer-unlimited-vacation.html, accessed April 13, 2021

[169]Stillman, Jessica, "How One Company's Unlimited Vacation Policy Totally Backfired", Inc. (magazine), July 14, 2015, https://www.inc.com/jessica-stillman/how-one-company-s-unlimited-vacation-policy-totally-backfired.html, accessed September 3, 2020

[170], North Carolina History Project, "Griggs v. Duke Power", John Locke Foundation, https://northcarolinahistory.org/encyclopedia/griggs-v-duke-power/, accessed August 27, 2020

[171]Legal Information Institute, "Smith v. City of Jackson [03-1160] 544 U.S. 228 (2005) 351 F.3d 183, affirmed.", https://www.law.cornell.edu/supct/html/03-1160.ZS.html, accessed August 28, 2020

[172]Preston, Eliza, "Opinion Recap: *Meacham v. Knolls Atomic Power Lab*", Supreme Court of the United States Blog, June 23, 2008, https://www.scotusblog.com/2008/06/opinion-recap-meacham-v-knolls-atomic-power-lab/, accessed August 28, 2020

[173]Smith, Allen, "House Passes Protecting Older Workers Against Discrimination Act", January 16, 2020, Society for Human Resource Management, https://www.shrm.org/resourcesandtools/legal-and-compliance/employment-law/pages/house-passes-protecting-older-workers-against-discrimination-act.aspx, accessed October 2, 2020

[174]Kim, Pauline T. and Sharion Scott, "Discrimination in Online Employment Recruiting", July 16, 2018, St. Louis University Law Journal Vol. 63 No. 1, https://papers.ssrn.com/sol3/papers.cfm?abstract_id=3214898, accessed November 3, 2020

[175]Terrell, Kenneth, "Facebook Reaches Settlement in Age Discrimination Lawsuits", March 20, 2019, https://www.aarp.org/work/working-at-50-plus/info-2019/facebook-settles-discrimination-lawsuits.html, accessed August 28, 2020

[176]Goss, Stephen C., "The Future Financial Status of the Social Security Program", United States Social Security Administration, Office of Retirement and Disability Policy, Social Security Bulletin, Vol. 70, No. 3, 2010, https://www.ssa.gov/policy/docs/ssb/v70n3/v70n3p111.html, accessed December 28, 2020

[177]National Association of Retailers and American Student Assistance, Student Loan Debt and Housing Report 2017, https://www.nar.realtor/sites/default/files/documents/2017-student-loan-debt-and-housing-09-26-2017.pdf, accessed December 28, 2020

[178]Leonhart, David "The American Dream, Quantified at Last", The New York Times, December 8, 2016, https://www.nytimes.com/2016/12/08/opinion/the-american-dream-quantified-at-last.html?, accessed September 3, 2020

[179]Pew Research Center, "Millennials in Adulthood", Chapter 2, March 7, 2004, http://www.pewsocialtrends.org/2014/03/07/chapter-2-generations-and-issues/#views-of-social-security

[180]Steuerle, C. Eugene and Caleb Quackenbush, "Social Security and Medicare Taxes and Benefits over a Lifetime: 2017 Update", June 5, 2018, https://www.urban.org/research/publication/social-security-and-medicare-lifetime-benefits-and-taxes-2017-update, accessed September 3, 2020

[181]Social Security Administration, "Social Security History, Ratio of Covered Workers to Beneficiaries", https://www.ssa.gov/history/ratios.html, accessed September 3, 2020

[182]Goodman, John C., "Why Was Social Security Designed Like A Ponzi Scheme?", Forbes Magazine, Aug 13, 2015, https://www.forbes.com/sites/johngoodman/2015/08/13/why-was-social-security-designed-like-a-ponzi-scheme/, accessed December 28, 2020

[183]Droblyen, Eric, "Are your 401(k) fees 'Reasonable'? Benchmark Them to Find Out", Employee Fiduciary [401(k) administrator] June 28, 2017, https://www.employeefiduciary.com/blog/are-your-401k-fees-reasonable-benchmark-them-to-find-out, accessed September 3, 2020

[184]Cilluffo, Anthony, "5 Facts About Student Loans", Pew Research Center, August 13, 2019, https://www.pewresearch.org/fact-tank/2019/08/13/facts-about-student-loans/, accessed September 4, 2020

[185]Corporate website of Johnson, Kendall & Johnson, https://jkj.com/our-services/retirement-services/, accessed May 24, 2020

[186]Kenton, Will "Monte Carlo Simulation", August 25, 2020, https://www.investopedia.com/terms/m/montecarlosimulation.asp, accessed September 4, 2020

[187]Clark, Joseph, "Ameriprise Agrees to Pay $27.5 Million to Settle Fiduciary Breach and Prohibited Transaction Claims", Proskauer [Law Firm], April 7, 2015, https://www.erisapracticecenter.com/2015/04/ameriprise-agrees-to-pay-27-5-million-to-settle-fiduciary-breach-and-prohibited-transaction-claims/, accessed September 4, 2020

[188]U.S. Securities and Exchange Commission, SEC Charges HSBC Securities for Misleading Retail Clients On Adviser Compensation", March 16, 2020, https://www.sec.gov/enforce/34-88387-s , accessed September 4, 2020

[189]Center for Medicare Services, 2020 Annual Report of the Boards of Trustees of the Federal Hospital Insurance and Federal Supplementary Medical Insurance Trust Funds, https://www.cms.gov/files/document/2020-medicare-trustees-report.pdf, accessed September 4, 2020

[190]Barry, Particia, "COBRA Bites Back", July 22, 2010, American Association of Retired Persons (AARP), https://www.aarp.org/health/medicare-insurance/info-07-2010/cobra_bites_back.html, accessed September 4, 2020

[191]Fidelity Investments, "How to plan for rising health care costs", August 3, 2020, https://www.fidelity.com/viewpoints/personal-finance/plan-for-rising-health-care-costs, accessed September 4, 2020

[192]Deming, David J., "The Growing Importance of Social Skills on the Labor Market", June 2017 (revised), https://www.nber.org/system/files/working_papers/w21473/w21473.pdf, accessed November 4, 2020

[193]Polk, Andrew, "China is Quietly Setting Global Standards", Bloomberg Opinion, May 6, 2018, https://www.bloomberg.com/opinion/articles/2018-05-06/china-is-quietly-setting-global-standards, accessed November 4, 2020

[194]Estes, Adam Clark, "U.S. Spies Can't Stop Buying Fake Microchips from China", The Atlantic, June 28, 2011, https://www.theatlantic.com/technology/archive/2011/06/us-military-fake-microchips-china/352255/, , accessed Novenber 4, 2020

ENDNOTES

[195]U.S. Attorney's Office for the Western District of Wisconsin, "Chinese Company Sinovel Wind Group Convicted of Theft of Trade Secrets", January 24, 2018, https://www.justice.gov/usao-wdwi/pr/chinese-company-sinovel-wind-group-convicted-theft-trade-secrets, accessed December 28, 2020

[196]Commission on the Theft of American Intellectual Property, hosted by the National Bureau of Asian Research, https://www.nbr.org/program/commission-on-the-theft-of-intellectual-property, accessed November 4, 2020

[197]Autor, David, David Dorn, Gordon Hanson, Gary P. Pisano, Pian Shu, "Competition from China reduced innovation in the US", Centre for Economic Policy Research, March 20, 2017, https://voxeu.org/article/competition-china-reduced-innovation-us, accessed November 5, 2020